Dedication

I would like to thank Gowri, my wife, for her support and Aditya Tejas, our son, for his encouragement during the writing of this book.

Contents

Preface

This book is the outcome of discussions I have had with numerous business leaders and project portfolio management practitioners on their need for a single reference that covers portfolio, program, and project management perspectives, highlighting how these concepts contribute to better business performance. Although many texts exist that specialize in each of the above perspectives, we felt that an integrated treatment at an overview level—incorporating the best practices from diverse global best standards, in a cohesive flow—would be valuable to business and practitioners alike.

My background experience of devising strategies for multiple companies (especially the Information Technology portfolio strategy) and deploying them through numerous programs and projects came quite handy in interpreting the project portfolio best practices. I have thus endeavored to condense the core elements of project portfolio management and put them together in a logical, end-to-end flow, so that businesses, practitioners, and academic institutions will gain valuable insights for practical real-life implementations.

I have also added separate chapters on transition management, change management, benefits management, and the Enterprise Project Management Office. I believe that the addition of these chapters can guide practitioners in better project portfolio implementations.

An illustrative case study, covering the application of project portfolio concepts to a practical scenario, has been provided to integrate and reinforce the topics. Templates that can be used by practitioners for their portfolio, programs, and projects have also been added towards the end of the book.

We believe that the right application of project portfolio concepts can immensely benefit both commercial and non-profit organizations in achieving their strategic objectives, delivering superior business performance and enhancing their professionalism. Towards this objective, we look forward to the assimilation and use of the best practices covered in this book by the business and project portfolio management community.

Acknowledgments

This book would not have taken shape without the active encouragement and guidance of Dr. Ginger Levin, who was instrumental in this venture. Dr. Levin went through the entire draft, suggested changes, and offered her unstinted support throughout. I would also like to extend my thanks to John Wyzalek for his support in publishing this book through CRC Press.

Marje Pollack, the Production Coordinator (DerryField Publishing Services), went through the proof edits meticulously and worked closely with me to get the book in good shape. I would like to offer my sincere thanks to her.

– October 2015
Bangalore, India

About the Author

Ramani has over 25 years of experience in the technology and management consulting industry, spanning project, program, and portfolio management; management consulting (with PwC Consulting); Information Technology strategy/portfolio development; and client-relationship management. He has successfully managed many large projects/programs relating to ERP implementations, business-systems integration, and IT strategy development. Prior to consulting for PwC, Ramani managed technology and application services for numerous clients, including non-profit organizations. He has also handled portfolio, program, and project management workshops for large transnational clients in multiple countries.

Ramani is currently managing his own company, GRT Consulting LLP, which specializes in project, program, and portfolio management related consulting and training. Ramani is among the very few global professionals who are accredited in project, program, and portfolio management certifications, from both PMI and from AXELOS frameworks. He blends these best practices in his coaching and consulting engagements, integrating them with the client industry context.

Ramani holds a post-Master's educational qualification in Computer Science. He also holds multiple credentials, including the PfMP®, PgMP®, PMP®, and PMI-RMP® from the Project Management Institute (USA). In addition, he has earned multiple practitioner certifications from AXELOS/APMG (UK), including in PRINCE2®, MSP® (for Program Management), MoP® (for Portfolio Management), M_o_R® (for Risk Management), P3O® (relating to setting up and implementing PMOs),

Change Management, MoV® (for Value Management), CHAMPS2® (for enterprise-wide transformation program implementation), PRINCE2 Agile, and Benefits Management.

– Ramani can be contacted at ramani@grt-consulting.com.

Chapter 1

Context for Change

1.1 Why Change?

No organization is impervious to change. Rather, the survival and growth of an organization is dependent on how it can cope with change. Adapting to change is a critical success factor, which differentiates robust organizations with those that "go under." And this postulate works for commercial organizations as well as for non-profit organizations.*

The triggers for change can come from multifarious factors. The organizational context becomes a significant factor in determining which factors bear more importance. To illustrate, governmental organizations are more

* In his blog (http://www.aei.org/publication/fortune-500-firms-in-1955-vs-2014-89-are-gone-and-were-all-better-off-because-of-that-dynamic-creative-destruction), Mark J. Perry states the point that only about 12% of the companies survived the 60 years, from 1955, in the Fortune 500 list. Others went out of existence, merged with others, or dropped out of the Fortune 500 list. And many of the companies who did not survive had failed to adapt to change. Similar studies by Deloitte's Center for the Edge (www.forbes.com/sites/stevedenning/2012/01/25/shift-index-2011-the-most-important-business-study-ever/) show that the average life expectancy of a Fortune 500 company has declined from around 75 years half a century ago to less than 15 years today, and it is heading toward 5 years, if nothing else is done. The pressure to adapt to change is ever more increasing.

prone to changes because of political factors, as compared to commercial organizations (which are for-profit entities). Breakthroughs in technology are more likely to impact commercial organizations, as compared to non-profit organizations. The changing competitive landscape can become the most critical factor for transformation in commercial organizations, which may not be as pertinent for non-profit organizations.

But the bottom line is that all organizations need to manage change, which is facilitated by change initiative management. "Change initiative management" collectively encompasses what needs to be done within portfolio, program, and project perspectives to address change.*

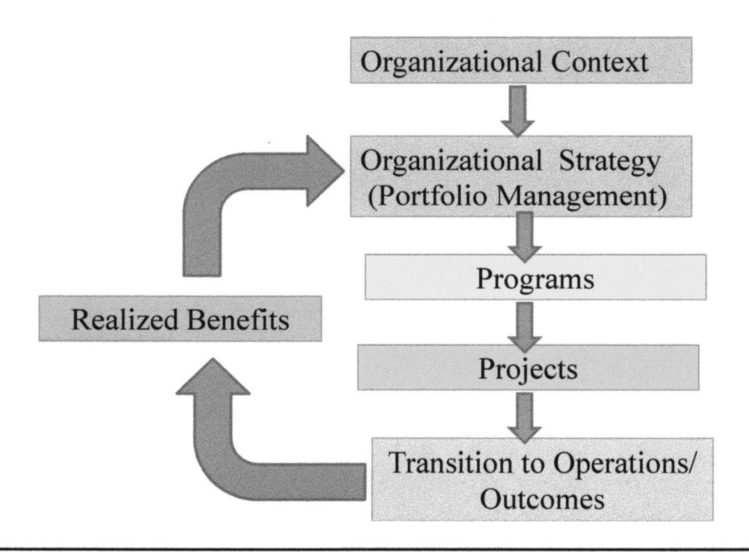

Figure 1.1 Change initiative management across multiple perspectives.

* "Change initiative management" is known by different terminologies as per varying organizations. The Project Management Institute (PMI) calls the change initiative management, "Organizational Project Management." As per the PMI, organizational project management is defined as the "Systematic management of portfolios, programs, projects in alignment with the organization's strategic business goals." (Quoted from the PMI standard: "Managing Change in Organizations: A Practice Guide"; http://www.pmi.org/~/media/Files/Home/ManagingChangeInOrganizations_A_Practice_Guide.ashx.). Many of the AXELOS frameworks refer to change initiative management as P3M (called "Portfolio, Programme and Project Management"), which is a crisp definition encompassing all three perspectives. This acronym is used in this book to connote change initiative management.

As we will see later in this book, all the above three perspectives are interrelated. Portfolio management decides which change initiatives need to be implemented as programs and projects (for the projects directly linked to the portfolio). Program management, in turn, spawns the projects that need to be executed.

Once the outputs from the projects are transitioned into operations (which are controlled by the functional departments), the outcomes and benefits are monitored, which can give a reverse feedback to strategy. This provides a way to determine if the programs and the projects taken up were indeed successful in achieving the strategic objectives of the organization.

A pictorial representation of how different perspectives of change initiative management are linked together is depicted in Figure 1.1.

As noted in this figure, the organizational strategy is impacted by the context in which it is positioned. For instance, not-for-profit organizations will have a different value chain, as compared to a for-profit organization.

1.2 Triggers for Change

The triggers for change can emanate through "political," economic, social, technology-driven, legislative, and environmental factors (usually known by the acronym PESTLE). The following illustrations indicate how change can be triggered by these six factors:

1. *Changes that are due to political factors.* For instance, change in the ruling party at the national level as a result of elections or change in the organizational structure can have ripple effects. When two companies merge or when a new Chief Executive Officer (CEO) joins, the expectations set off changes, which can be ascribed as ones that are due to "political" factors. In many countries, which party comes into power can influence the overall direction of business for many companies.
2. *Changes that are due to economic factors.* The economic downturn during 2008–2009 "wiped" out the fortunes of many financial institutions. Banks in particular had to be restructured and had to undergo "stress tests" to prove their viability. This led to a redesign of their portfolio and the business lines they operate.

3. *Changes that are due to social factors.* These can be triggered because of migrations, increase in literacy levels, restructuring of social hierarchies, etc.
4. *Changes that are due to technology factors.* These changes are among the easiest to discern. Technology has played a critical role in facilitating social networks. Advances in mobile and cloud computing, data analytics, etc., could be quoted as examples, giving rise to new change initiatives for organizational survival, and on how they deal with customers.
5. *Changes that are due to legislative factors.* These changes, more often than not, have been seen in industries that are regulated, such as financial institutions, healthcare organizations, etc.
6. *Changes that are due to environmental factors.* These changes could be attributed to the concern for protecting natural assets and increasing the quality of life.

Market transparency, labor mobility, global investments, and instantaneous tools for communication have intensified global competition. For most of the global companies, moving forward has become embedded as a part of the work culture, as these companies appreciate that improvements and churns are brought about only by change. At the strategic level, organizations would be monitoring the triggers for change, along with customer feedback. In addition, successful organizations also involve critical stakeholders as a part of the change process.

Change initiative management enables managing complex change, which involves process, organization, and technology dimensions. However, most of the change initiatives fail because of inadequate involvement on the part of the concerned stakeholders. It is human nature to resist change and not to come out of one's comfort zone. This aspect will be discussed in a subsequent chapter (see Chapter 6: Change Management).

One critical aspect of change initiative management is that the rate of change itself is increasing, and the organizations of tomorrow will need to cope with more rapid changes. In this context, it would be useful to consider the concept of "velocity of change" of various industries. By their very nature, change in some of the industries is relatively slow moving, such as in the lumber industry. Change in other industries, however, is relatively fast moving, such as in information technology (IT) and telecommunications.

In a nutshell, whenever the strategic objectives of an organization change, a review of its collection of change initiatives is required. Successful organizations do not shy away from change but, rather, welcome it as an opportunity to "reinvent" themselves, discarding irrelevant practices and structures in the context of new settings.

1.3 The Impact of Change

The impact of change itself can vary across the perspectives. Portfolio Managers need to be open to changes coming from external factors, whereas Project Managers typically need to control changes within defined dimensions of scope, schedule, quality, and budget. Program Managers need to consider the top-down changes coming in from portfolio management as well as changes coming from the execution of projects and the resultant outcomes in operations.

In subsequent chapters, we expand on how the successful management of portfolios, programs, and projects facilitate coping with change.

Chapter 2

It All Commences with Strategy! Project Portfolio Management

2.1 Starting Point for Portfolio Definition

In Chapter 1, we saw that all organizations need to address change. Change could be entropic or could lead to a better order of things. Organizations respond to change by redefining their strategy.

A rational first step to take while redefining strategy is to assess what the context of change is and where the organization currently stands. This "as-is" assessment is easier for an ongoing business, as it has past records and plans (such as sales data, personnel productivity information, etc.) to use. For a start-up organization, further market analysis needs to be undertaken to ascertain where the organization currently stands, with reference to the industry and the competitive landscape.

One simple tool that can be used to commence this analysis is the SWOT matrix formulation. SWOT is an acronym that stands for "Strengths, Weaknesses, Opportunities, and Threats" and is widely used in management.

To give a simple illustration for SWOT, consider a scenario in which the new government of a country (which is dependent on the continued support of its coalition partners) has reformulated the regulations for setting up new banks, to have a wider banking reach to its underprivileged population. For a prospective bank considering entry, the following SWOT matrix (see Figure 2.1) is applicable.

In this matrix, two entries have been noted for each grid (in reality, there would be more entries, arising out of brainstorming discussions, etc.). It can be noted that whereas strengths and weaknesses are internal to the organization, threats and opportunities are external.

It is also quite likely that competitors perceive similar threats and opportunities, but the differentiation occurs during the recognition of the strengths and weaknesses unique to the particular organization and acting on this information.

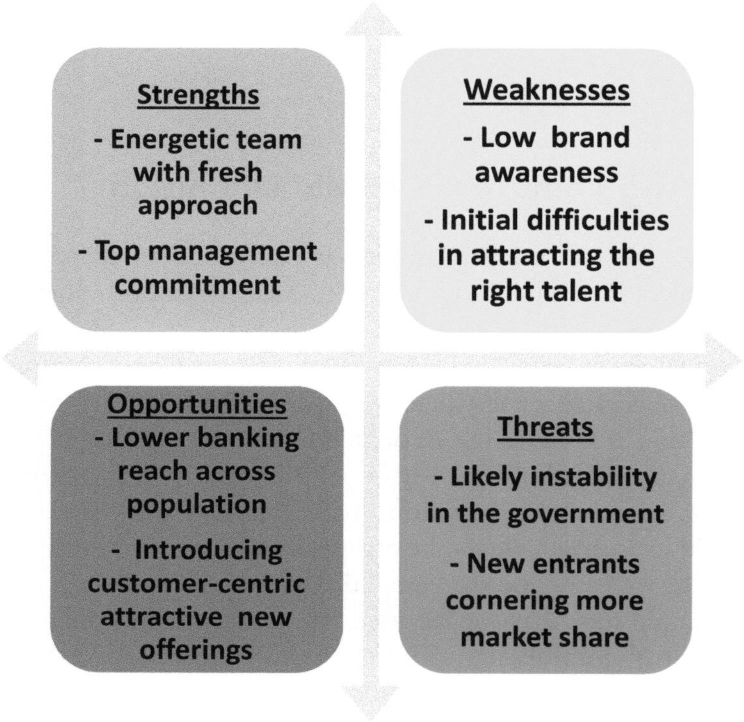

Figure 2.1 Illustrative SWOT matrix.

2.2 Strategic Positioning of Organizations

As a part of strategy formulation, the organization needs to decide where it needs to position itself during, say, the next five years. This time horizon can be dependent on the "velocity" of the industry referred to in Chapter 1 and the context and expected dynamic nature of changes. Many companies maintain a longer time horizon of, for example, five years, together with a mid-term perspective of three years, and they keep refining the operational plans for a year ahead.

The strategic objectives of the organization are set next, depending on where the organization is currently positioned and where it wants to move during the planning horizon. Usually these objectives are set in financial and market-facing dimensions for easier understanding. Expectations of the shareholders, top management, and funding organizations (if applicable) are taken into account while setting these targets.

To illustrate, a leading Enterprise Resource Planning (ERP) software provider has stated that its key performance indicators will be focused on growth, customer profitability, and employee engagement. Growth performance targets have been set for the coming year (2015), three years forward (for the year 2018), and a five-year outlook (for the year 2020).

Usually these goals are defined at stretch target levels to motivate diverse stakeholders. These company-level targets are split across Strategic Business Units (SBUs) or at the functional levels. This disaggregation calls for understanding the performance of the SBUs in the past and their perceptions of the market for future.

2.3 Boston Consulting Group (BCG) Matrix—Application

During this analysis, tools such as the Boston Consulting Group (BCG) matrix[1] are immensely useful.

As an illustration of the BCG matrix, the organization can group its products in grids representing the current market share of the product and the expected market growth.

In Figure 2.2, the X axis represents the current market share of the organization for the product, and the Y axis represents likely market growth in the future for the product. It may be noted that, although

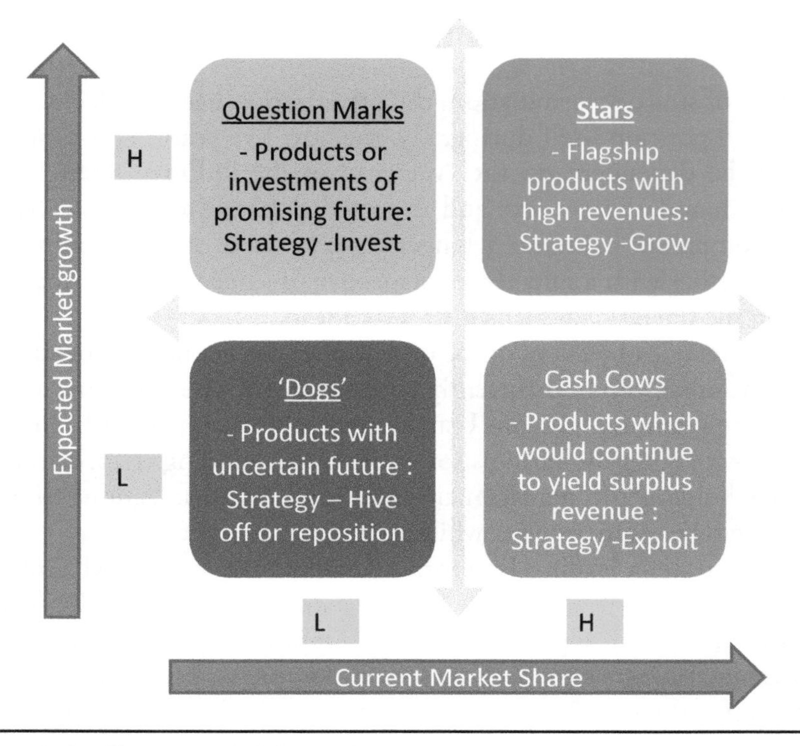

Figure 2.2 Illustrative BCG matrix.

products have been taken for illustration, the reference could also relate to geographies or SBUs.

In this context, "Question Marks" relate to those products with a current low market share but with a promising expected market growth in the future. They could relate to, for instance, breakthrough technologies (as in cloud computing, 3D printing, etc.) or new geographies, which could be "make or break" investments for the company. Because of this uncertain nature, investments herein are usually denoted as "Question Marks." Typically, the change initiatives undertaken for the products/markets in this quadrant could relate to research and development projects, feasibility studies, proof of concept surveys, etc.

Some of the "Question Marks"—once they get "proven"—move on to the "Stars" quadrant. Usually, they relate to "flagship" products, which are high-revenue earners for the company. Such products may require considerable marketing and upgrading efforts to keep up with customer expectations and the actions of competitors to keep up their market share.

Because of market forces, changes in customer expectations, changed regulations, and the entry of new competitors, flagship products degenerate into "Cash Cows." Such products normally require more of a maintenance type of investment, as compared to upgrading efforts. However, these products usually yield surplus revenues that are due to their already installed base and past popularity. An example could be a few of the applications hosted by Information Technology (IT) companies on mainframe computers for specific verticals, etc.

Ultimately, the "Cash Cows" degenerate in the market space because of obsolete technology and changes in customer preferences. Such products having a low market share and a bleak expected market growth in the future are termed "Dogs or pets." These products (or markets) need to be hived off or reoriented to become "Question Marks," with additional innovations. One illustration could be companies producing fax machines, which need to reorient their products since these machines are not widely used anymore.

This cycle mirrors the product lifecycle and can relate to the share of investments going in for diverse products or services being offered by the organization, which need different skill sets for different quadrants.

2.4 Setting Up of Performance Targets

The organization sets the performance targets to be attained during the reference period. While understanding the context for change, it would have also understood the current "as-is" situation, including the current profile of the customers, revenue streams, and concern areas for improvement through SWOT analysis, etc.

Target setting has always been a contentious area falling under the realm of management. Some of the publicly listed companies are driven by shareholder and financing institutions' expectations for return. Nonprofit organizations can set service-level targets. Usually, the inputs from a SWOT analysis, the BCG matrix, or a value chain analysis highlight critical issues (usually limited to five or six) that require immediate management attention. These issues (usually called "must address imperatives") can also facilitate setting targets, along with benchmarking with peers in industry.

The "incremental setting of targets" based on last year's performance (e.g., grow by 5% this year) is usually fallacious, as it does not take into account the industry growth rates. Some organizations adopt "zero-based" budgeting, wherein each year justification for each of the initiatives is considered afresh, rather than allowing them to continue uncontested. Whereas the advantage of this approach is eliminating unviable initiatives at the outset, the disadvantages could relate to having too much "churn" in the portfolio, initiatives getting "booted out" because initial benefits were unconvincing, and having the need to make too many adjustments and analyses during the commencement of each business change lifecycle. The durations of these lifecycles can vary across various portfolio categories and are also based on the industry velocity change referred to in Chapter 1.

It is always necessary to have baseline data comprising the current values of the metrics to understand the gap. The organization defines the Key Performance Indicators (KPIs), which are usually lead metrics and can be measured based on the data emanating from multiple sources. These KPIs should be able to be updated on a periodic basis and preferably linked to the individuals' performance objectives. In case the balanced scorecard approach (which will be explained later in this chapter) is applied, financial targets are set first and cascaded down to the customer, internal process improvement, and, finally, to learning and growth perspective targets. Apart from enabling a "cause and effect" linkage, the setting of financial targets first connects well with key stakeholders. Thereafter, when targets for the subsequent perspectives are set, they are seen to be facilitating the achievement of financial targets, rather than creating "pushback effects." This "dependency analysis" is imperative to understand the linkages across the initiatives, which may also give rise to risks.

Accompanying analysis, such as root cause analysis, can also provide inputs for disaggregating the overall target across the SBUs or geography levels. Many companies resort to tools such as "Scenario Analysis,"[2] considering diverse possibilities—as in worst case, normal case, and worst case for the influencing variables, while considering fixing targets.

2.5 Strategy Evolution

Once the targets are set, the organization needs to evolve the strategy, which could be defined as a coherent group of initiatives undertaken to achieve the targets. In the above definition, "group" indicates a collection of initiatives, whereby the integrated whole produces more value, as compared to those initiatives taken up individually. It should be noted that strategy is unique to the organization based on the context and more attuned to its internal strengths and addressing the weaknesses. Assuming all the competitors perceive similar opportunities and threats for an industry, the addressing of strengths and opportunities differentiates one company from another. Since the strategy is always defined after considering the context, the changing landscape ought to have been factored into its definition. In addition, since the future cannot be predicted accurately, the strategy can (and should) be amenable to change, based on reverse feedback.

It is quite imperative that the organizational targets (with SBU or geography or product-level breakup) need to be agreed upon across the leadership team. Usually, many companies commission external consultants to facilitate this process to bring in an element of objectivity.

2.6 Organizational Vision, Mission, and Strategic Objectives

Most of the reputed companies have a well-defined vision, mission, and value systems that are well published. Typically, a vision statement indicates a future state of the company based on the end goal it aspires to achieve.* Usually, vision statements do not change drastically; otherwise, it could confuse the major stakeholders. It also needs to be taken into account that the vision statement needs to be amenable to quantification. One well-known financial services company simply defined its vision statement as "to double the revenue in the next five years," which provides a succinct and clear goal that it intends to achieve. Simply stated, the vision statement indicates where the company is going and how it wants to be seen in the future. The vision statement should be sufficiently motivating to key stakeholders, communicated widely, and convey the reason why the organization cannot stay in the current state. Many change leaders create

* For instance, Toyota Corporation's vision statement commences with the following paragraph: "Toyota will lead the way to the future of mobility, enriching lives around the world with the safest and most responsible ways of moving people. Through our commitment to quality, constant innovation and respect for the planet, we aim to exceed expectations and be rewarded with a smile. We will meet our challenging goals by engaging the talent and passion of people, who believe there is always a better way." (Available from http://www.toyota-global.com/company/vision_philosophy/toyota_global_vision_2020.html)

the agenda for change by using the metaphor of a "burning platform" indicating that a drastic change in direction is a must, else the company could be "doomed." It is quite likely that this sense of urgency is amplified by feedback from customers, competitor actions, and impending change in external triggers—especially changes of a political or legislative nature.

The mission statement goes deeper into the purpose of the company's existence and what it does. It is the mission statement, which clearly points to the strategic positioning of the company.

Many companies also add "value systems" to connote what they believe in and stand for as part of the mission statement, and they enumerate the "Corporate Social Responsibilities (CSR)" that incorporate the companies' value systems, as well.

2.7 Environmental Scanning and Competitive Strategies

Environmental scanning is a major step used by companies while formulating strategy.

Along with the SWOT analysis, Porter's "Five Forces" framework[3] provides a useful guideline during the scanning. As per this framework, the intensity of competition being faced by the organization depends on the bargaining power of suppliers and buyers, barriers to entry, and the threat of substitute products. It has to be noted that the intensity of forces can vary across industries, and this model needs to be used along with other data-points—for example, profitability analysis of comparative firms.

The strategy of the organization needs to be clear regarding the following queries:

- What are the geographies and who are the target customers to be serviced?
- Which products/services will be delivered to them and how?
- What do we need to have in our organization to deliver them—in terms of capability, capacity, and skill sets?
- How do we know that our strategy is working?
- Which change initiatives do we need to take to acquire the above?

Portfolio Management seeks to address all the above queries at an organizational (or at an SBU) level.

The value proposition of the company highlights the "Unique Selling Proposition," which differentiates it from others and positions it in the competitive landscape. Porter's "Competitive Strategy"[4,5] highlights the following "generic strategies," adopted by the companies as value proposition statements.

- Cost competitiveness—selling at a cost lower than the competitors (e.g., Walmart)
- Product differentiation/innovation—having features that are valued by the customers (e.g., Apple)
- Focus—operating in a niche segment, possibly with entry barriers and long gestation periods (e.g., high-end pharmaceutical companies with niche products)
- Customer lock-in—because of tight supply chain management, system switching costs, etc. (some of the suppliers in the automotive and utilities/oil and gas environment)

In order to determine which generic strategy may work best, companies usually undertake their "value chain" analysis.[6] Value chain is the difference between the value (or revenue) the company gets from its customers less the "cost" of performing all its activities. Such activities get grouped under two major categories—"core" activities (including design, manufacturing/service creation, distribution, marketing, delivery, and maintenance) and "support" activities (including human resources, procurement, IT infrastructure, etc., which are usually shared functions across the organization). The degree of focus on these activities could vary across the companies based on the value proposition and the industry in which they are positioned. To give an illustration, an aircraft manufacturing company may focus more on design aspects, as compared to a fast-moving consumer goods company, which may concentrate more on supply chain and marketing aspects.

2.8 Application of Balanced Scorecard (BSC) to Portfolio Management

When it comes to strategy implementation, the "Balanced Scorecard" is among the most preferred tools used by most companies. The balanced scorecard is used extensively in business and industry, government, and

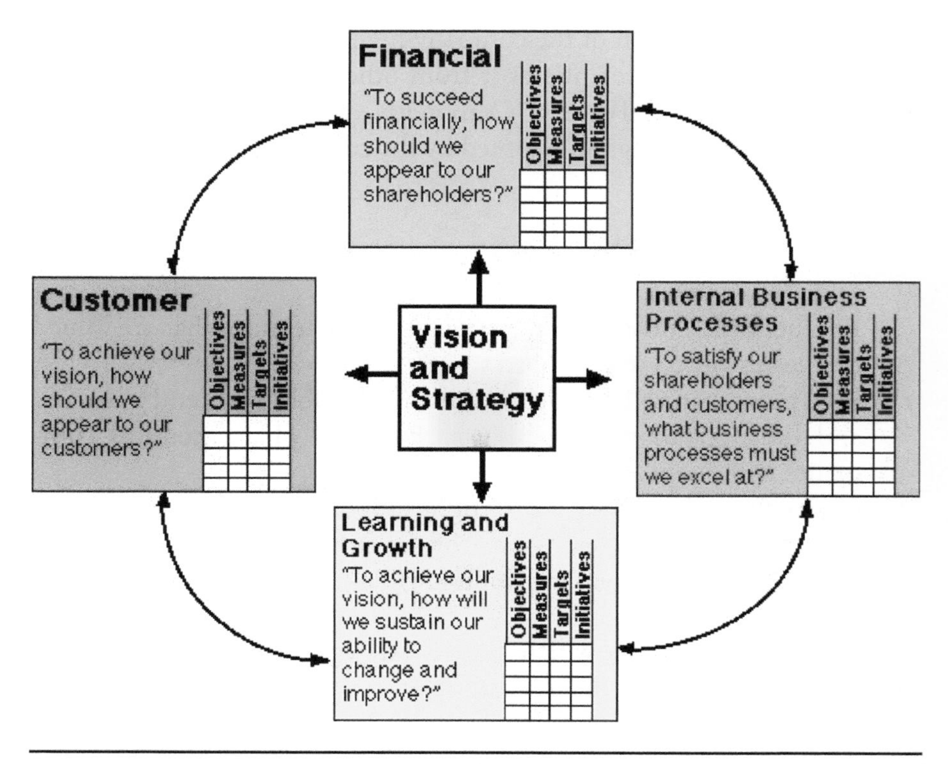

Figure 2.3 Balanced Scorecard—an overview.

nonprofit organizations worldwide. It facilitates the alignment of the organizational portfolio with the Corporate vision and strategy. Further, it enables tracking of the organizational performance against its strategic objectives. Internal and external stakeholder engagement is also facilitated by having such a robust performance management system.

The balanced scorecard[7] has four perspectives—financial, customer, internal process improvement, and learning and growth—as seen in Figure 2.3. Each of these perspectives has associated objectives, metrics to measure performance, goals (targets), and change initiatives to be taken to bridge the gap. Collectively, all the change initiatives in an organization would represent the organizational portfolio.

Descriptions of these four perspectives are as follows:

a. The **financial perspective** determines which factors appeal to the shareholders the most. The metrics to measure this perspective can include cash flow, sales growth, operating income, return on equity, etc. Traditionally, most companies were measured by their performance on these measures, and, hence, the internal short-term focus was to maximize them. In most companies, the finance department is vested with the responsibility of maintaining these metrics, which are lag indicators that measure the tangible outcomes from the execution of the strategy. The choices to be made here by the top management typically consist of whether the company is looking at boosting up its share price by declaring dividends now or deferring the payouts for longer-term gains by reinvesting in growth.

b. The **customer perspective** addresses the question of how customers see an organization, and, hence, it defines the source of value. The metrics to measure performance in this perspective can include the Customer Satisfaction (CSAT) Index, the voice of the customers, new customers acquired, the percentage of repeat sales, and more. The marketing and sales departments in organizations are usually entrusted to monitor these metrics and take corrective actions as needed. Initiatives in this area also address which customers to serve, which geographies to target, and what types of products and services will be provided to them.

c. The **internal process perspective** seeks to address the question regarding what companies need to do to excel in the marketplace and to enhance performance in the organization itself. The metrics to measure performance could include, for example, process rationalization, business process reengineering, tighter supply chain management, and total quality management. This perspective can be owned by multiple functional departments—such as quality, the Center of Excellence (COE)—in the organization, which differentiate its performance. Consistent with the customer perspective, the internal process perspective will determine if the company can serve its customers to retain superiority in the marketplace.

d. The **learning and growth (or development) perspective** looks at the question of how companies create a learning organization, enhance its skill sets, continue to improve, and innovate. This perspective can extend to include the development of robust internal processes for new product development, information technology support, and more, and generally create intangible assets. Again, consistent with the earlier three perspectives, the learning and growth perspective looks at competency enhancement to use the redefined processes to address customer requirements in the required markets. In a way, the balanced scorecard provides a framework that not only focuses on the financial metrics but also balances efforts and initiatives across the other three perspectives to drive the financial success, which can collectively form the composite portfolio of the company.

2.9 Balancing the Portfolio

Companies with a longer vision initiate changes in all four perspectives for balanced and sustained growth. For instance, for a pharmaceutical company, immediate revenues can come from the sales of established products, whereas future revenue streams can come only through investments in research and development, creation of knowledge databases/patents and increased collaboration. The lead indicators associated with these perspectives provide early evidence regarding achievement of the targets concerning customer facing and financial perspectives. The strategic initiatives are taken to address the performance target gaps, which together can collectively form the organization-level portfolio.

It should be noted that companies need to balance their efforts in strategy planning and execution. Many companies are competent in developing a strategy, as it's mostly done top-down from "ivory tower," and it provides intellectual stimulation and a "feeling of euphoria" at being able to set the path for growth. However, the real challenge comes in its execution. Because of poor strategy or alignment with vision, or a lack of buy-in from stakeholders, or a combination thereof, strategy execution can flounder. Some companies move into execution without adequate planning, which calls for frequent revisions in strategy. Neither of these is desirable, and a healthy combination of strategy planning with reverse feedback and a "closed-loop" system, which enables the strategy to be refined based on operational performance, is most ideal.

2.10 Portfolio Definition and Management—Roles and Responsibilities

Clearly, a portfolio definition cannot only be done "top-down." Multiple skill sets are required to define the portfolio and manage it. We define herein key roles that can be used during change initiative management. Obviously, different companies give varying nomenclatures to the terms specified here, and tailoring of the respective roles can be implemented accordingly.

The highest-level team, which is required in portfolio management, can be called Portfolio Steering Group (PSG). Usually the PSG consists of senior leaders who drive the strategy and take ownership for the success of the portfolio. This Group is responsible for investing in the portfolio and approving its final composition. The governance frameworks relating to portfolio management itself need to be approved by this Group. Since this Group becomes accountable for the portfolio, its members need to ensure that the portfolio is suitably balanced, and that resources are allocated appropriately.

The Portfolio Office (PfO) (or similar entity) is responsible for defining (or refining) the portfolio. This Office is responsible for finalizing the composition of the portfolio and getting it approved by the PSG, by prioritizing strategic initiatives, and, by consulting key stakeholders (especially the middle managers and the functional units) so that the "buy-in" for the portfolio becomes easier. Extensive consultations and nominal group techniques can be used by the Portfolio Office (along with the input from Subject Matter Experts) to facilitate portfolio definition.

The Portfolio Sponsor usually heads the Portfolio Office and is also represented in the PSG. The Sponsor is a top-level manager, and in some companies is also known by the term Portfolio Director. The Portfolio Manager, who is responsible for coordinating the efficient operations of the portfolio, reports to the Portfolio Sponsor. The Portfolio Manager takes the assistance of the Portfolio Office and is vested with the responsibilities of facilitating the refining of the portfolio as needed.

Portfolio progress is tracked by the Portfolio Progress Monitoring Group (PMG). This group can consist of key functional heads and business leaders to monitor the achievement of strategic objectives and make recommendations to the PSG if any mid-course corrections are needed on an ongoing basis. The portfolio dashboard reports for consideration by the PMG can be prepared by the Portfolio Manager/Portfolio Office. The PMG is also usually headed by the Portfolio Sponsor, to provide a linking pin to the Portfolio Steering Group.

Finally, in order to maintain data and best practices/lessons, a COE can be attached to the Portfolio Office (PfO). It may be noted that the PfO is a standing entity, which can also render scrutiny and challenge to the decisions (including those taken by the senior management team on the composition of the portfolio). During the process of setting the targets, not only are the issues considered, but opportunities are reviewed as well. The organization can use multiple tools/techniques applicable across the balanced scorecard perspectives during change initiative planning and execution. Some of these tools can include:

- Financial perspective: Activity-based costing
- Customer perspective: Customer relationship management
- Internal process perspective: Six Sigma. Lean, Total Quality Management, International Standards Organization (ISO), etc.
- Learning and growth perspective: Individual and team change management/organizational change management

The most effective scorecards contain both lead and lag metrics. Usually, the lead metrics relate to the outcomes (such as time spent with a new prospect), and the lag metrics relate to the benefits (such as revenue generated from new accounts). Lead metrics are usually assigned to the internal process and learning and growth perspectives. Lag metrics are usually associated with the customer perspective and, more often than not, to the financial perspective. It should also be noted that when the performance is measured by the lag metric, such as a benefit (which could be realized much later), a couple of lead indicators are included to ensure that the initiative is itself in the right trajectory.

The portfolio management definition takes into account the existing change initiatives and new initiatives proposed to be taken to achieve the targets. The PfO works through the capability and capacity constraints to define a viable portfolio for the organization.

As an illustration, a telecom major wanted to increase the Customer Satisfaction (CSAT) for its services from key customers. As per the studies it commissioned, the baseline CSAT value was measured as 4.1 (out of 5.0). The company wanted to increase the CSAT score to 4.5 over the next three years. To achieve this objective, it identified multiple initiatives (e.g., customer segmentation and focused coverage) to address this gap. Each of the change initiatives became a program (or a project), collectively forming the portfolio of the company. While determining the set of initiatives, the company considered "in-flight" initiatives and understood the existing gaps while defining the composite portfolio.

Once the objectives were formed, the PfO mapped which initiatives addressed which objectives, and if there were gaps and overlaps in coverage. During this mapping, the following aspects were considered:

a. Which are the ongoing initiatives that do not address any strategic objectives?
b. What strategic objectives are addressed by multiple initiatives as duplicates?
c. What strategic objectives did not have linked initiatives, so they could be added?

It should be noted that this rationalization can typically lead to stakeholder engagement issues, so the direction and guidance from the PSG was required. The initiatives were approved after a Business Case viability analysis so that clear accountability was fixed, and progress could be measured. While mapping the project outputs to the benefits, only clear linkages were considered. Otherwise, spurious linkages could have led to the justification of "pet projects," especially ones sponsored by Senior Managers to justify their existence. Likewise, the expected benefits projected in the business cases were validated critically.

It has generally been the tendency in some companies to overstate the benefits and understate the risks to gain approval or to be overly optimistic in their benefit projections because of exuberance not backed by reality. Many of the companies that were launched and then folded during the dot.com boom and bust from 1995 to 2001 bear testimony to this predisposition.

In order to redesign its portfolio, the telecom major reviewed its vision, mission, and strategic objectives. It then set targets to be achieved for major strategic objectives for the defined planning horizon. From an analysis of the "as-is" situation, the organization baselined the current values for these objectives and finalized the change initiatives to bridge the gap between targets and current baseline values. Balanced scorecard and the BCG matrix were useful tools to finalize the collection of initiatives. Thus, portfolio management focuses on "effectiveness" (or are we doing the right initiatives) as contrasted to "efficiency" (are we doing the things right).

The work done under portfolio management can thus be summarized as follows:

a. Identifying, prioritizing, and aligning the initiatives with the organizational strategic objectives
b. Funding the initiatives and allocating resources, as required, to get the maximum return on investment
c. Putting in place the governance systems that are to be applicable for strategic initiatives
d. Assessing the portfolio value and reprioritizing the initiatives, as needed, based on phase/gate reviews
e. Managing critical risks and communicating the portfolio performance to top management and to key stakeholders

Portfolio management is essential for an organization to conserve its critical resources and put them to optimal use. Initiative managers need to get clear guidance on how to manage their change initiatives, which is provided by the portfolio management. Every organization should also consider its capability/capacity in terms of skill sets and overall risk appetite, while adding new change initiatives to the portfolio. Thus the strategy sets the "canvas" for the portfolio management, whereas the implementation of the portfolio provides a reverse feedback if the strategic objectives are being achieved.

It is not mandatory to attain a high level of maturity and capability in project and program management before embarking on portfolio management. This is a fallacy held by many organizations as they continue to hone the project and program management competencies, while not reaping the benefits.

The primary question to be answered here is, "Are we investing in the right change initiatives?" which is addressed by the portfolio management. Conversely, having an effective portfolio management does not preclude the need for successful project and program management practices. Portfolio management pays for itself by putting in processes to guide initiative implementation effectively. More than that, a robust portfolio management system rules out poorly conceived projects and programs being taken up or continued, thus conserving critical organizational resources. Successful implementation of portfolio management also depends on the culture of the organization (including openness to change or discipline in execution) and the scale at which it is implemented. Some organizations try the portfolio management approach for smaller SBUs and, after learning the lessons in the process, move on to implement it in other SBUs.

The COE in the Portfolio Office can be a critical entity here in collating lessons learned and providing reverse feedback during implementation.

2.11 Portfolio Definition and Implementation—Key Steps

Let us now look at the key steps that need to be taken for portfolio finalization and implementation. It is assumed that the organization has a well-defined and an agreed-upon vision, mission, and strategic objectives as well as an initial list of change initiatives formulated (possibly identified through the balanced scorecard approach). Most portfolios are not implemented in a "green-field" environment, and there are ongoing initiatives that need to be considered while defining or refining the portfolio.

The following major steps are considered during portfolio finalization:

a. **Collect:** A major first step is to collect information on the ongoing change initiatives and classify them. Over the course of years, many organizations have been running projects that have become partly irrelevant but continue to run because no one questioned (or care to question) why they should be continued. The information collected herein can relate to the scope, schedule, and budgets; resources assigned to the initiative; current performance status; and risks and benefits. This process can be daunting, especially for large companies with many ongoing change initiatives. The PfO can facilitate creating the portfolio register. Unless this information is reliably collected, it would be difficult to proceed further towards prioritization. Otherwise, resources would seemingly be deployed in ongoing initiatives that add no incremental value to the organization, and such resources need to be available for more critical initiatives.

   During the enumeration, it may be too overwhelming to list all the change initiatives ongoing in a large organization, unless the organization is considering a total capacity management model. Typically, it is useful to consider only those change initiatives crossing a defined threshold investment or specific initiatives, which are mandatory in nature, or crossing a set risk threshold. The thresholds need to be defined by the PSG and applied by the PfO during portfolio finalization.

Mapping these initiatives to strategic objectives becomes an "eye-opener" for the executive managers on the effectiveness of existing resource utilization and furnishing information on which initiatives need to be terminated or merged with others to conserve resources. In some organizations that run multilayered portfolios, this mapping can be done bottom-up and rolled upward.

The information to be collected can include the details of the name of the change initiative, the current owner/department, scope/schedule/cost, risk-related details, business case, current status (how many gateways have been completed), and information on strategic alignment/benefits expected from the initiative.

b. **Classify:** Once the set of initiatives has been collected, these initiatives need to be segmented across multiple "groups." Reference to the BCG matrix or the balanced scorecard here could be useful for classification and in determining further funding allocations for different grids or perspectives. As the initiatives are being classified, their interdependencies, if any, are also noted (some of this information could have been noted during the "Collect" step). The overall guidance for classification can be set by the PSG.

If the classification criteria are linked to the strategic objectives of the organization, it would be easier to ascertain which change initiatives support which strategic objectives and which objectives are not supported by any of the initiatives. The "necessary and sufficiency" criteria, widely used in disciplines such as mathematics and logical inference, can be useful here.

The "bucketing" or segmenting of the initiatives can facilitate better comparison. For instance, these groupings can relate to, for example, mandatory projects, research and development projects, maintenance projects, and turnaround projects. This grouping also facilitates ensuring that projects and programs of a lesser budget (but having more strategic value to the organization) do not get left out in comparison with larger-value initiatives.

c. **Evaluate and prioritize:** Once the change initiatives are grouped into multiple segments, they are evaluated and prioritized for further work. During this step, the application of weighting and ranking procedures is useful. The weighting and ranking procedure needs to be approved by the PSG to avoid any perceived bias during initiative prioritization.

The inputs for the prioritization can be provided by the PfO, which can also determine the initial composition for approval by the PSG.

The evaluation criteria for the change initiatives can change across the categories. For commercially oriented projects, metrics such as the Internal Rate of Return (IRR), the Net Present Value (NPV), and the Accounting Rate of Return (ARR), can be used to evaluate their attractiveness. In such cases, only those initiatives that cross a particular IRR or NPV threshold value are considered for further investment analysis. One of the usual fallacies we encounter during benefits evaluation is double counting. For instance, two or more change initiatives count the same benefit twice in their business cases to facilitate obtaining approval. Many companies have a centrally defined Benefits Manager role (who is more of a Subject Matter Expert) to address this double counting.

Once the likely returns and risks are assessed, the organization (or the Portfolio Office) can prepare a "bubble chart," which represents the level of risk and the likely returns of various proposed initiatives. This chart is also known as an "attractiveness/achievability" chart. Organizations also determine the risk threshold, above which they may not like to invest.

For instance, consider the following bubble chart (see Figure 2.4) for the risk-return matrix. In this matrix, the X axis represents expected returns (in %—based on metrics such as IRR, etc.), and the Y axis represents the likely total risk exposure of the initiative. The PSG usually sets the threshold levels of acceptability of risk and minimum threshold levels to deem an investment attractive. Thus, for illustration, any initiative with an expected return less than 15% is considered not feasible by top management. Likewise, any initiative having a total exposure crossing "Medium" is also unacceptable, as per the risk appetite thresholds set. Various change initiatives being considered are plotted, with the size of the bubbles representing their proposed investments.

As per the illustration depicted in Figure 2.4, only two preferred initiatives should be considered for further analysis. It should be taken into account that the risk thresholds could be set differently for different segments (e.g., R&D-related projects can have a higher risk threshold as compared to maintenance types of projects).

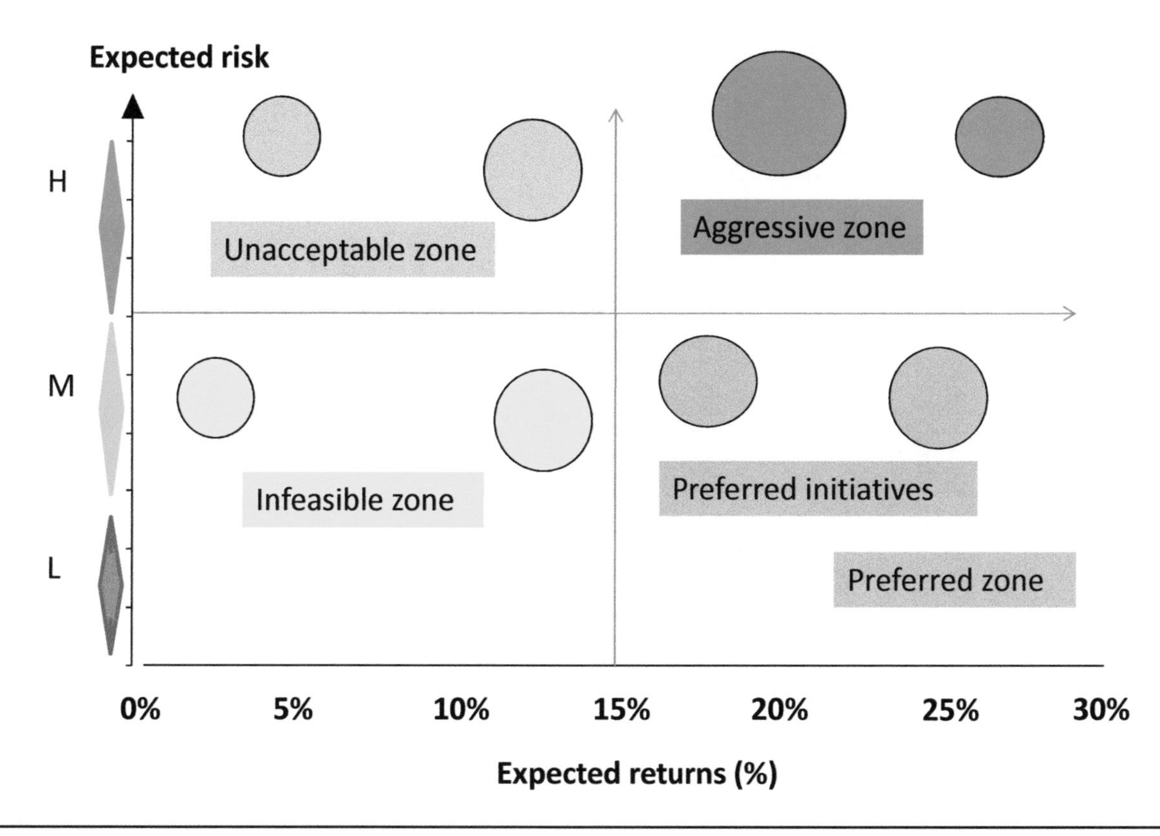

Figure 2.4 Risk return matrix.

Further analysis of initiatives to be shortlisted can depend on decision-making criteria, including pair-wise ranking or weighted prioritization algorithms. This sequence is broadly in line with the Analytic Hierarchy Process (AHP)[8] process followed by many companies for portfolio definition.

To give a simple example of the pair-wise ranking technique, consider the table depicted in Figure 2.5. In this table, five initiatives (initiatives A to E) are considered. If an initiative (such as A) is deemed to be better as compared to initiative B, then a score of 1 is assigned to the row corresponding to initiative A against the column of initiative B (and a score of 0 is assigned to the initiative B row against the column pertaining to initiative A). It has to be noted that these comparisons need to be done objectively by a group of stakeholders, facilitated by the PfO. Once all the pair-wise comparisons are completed, the row-wise totals are computed, and the initiatives with the maximum row total are considered further for investment. As per this comparison in the illustration, the total score for initiative A is 3, making it the most attractive initiative for investment. Initiatives B, D, and E are next with the scores of 2 each, and these initiatives could be next to be considered for selection.

In a weighted prioritization model, weights are assigned to various criteria (preferably adding up to 100%). Then, every initiative is scored for each of the criteria, and a weighted score is calculated.

	Initiative A	Initiative B	Initiative C	Initiative D	Initiative E	Row Total
Initiative A	-	1	0	1	1	3
Initiative B	0	-	1	0	1	2
Initiative C	1	0	-	0	0	1
Initiative D	0	1	1	-	0	2
Initiative E	0	0	1	1	-	2

Figure 2.5 An example of pair-wise ranking of initiatives.

		Initiative scores			
	Weight (%)	Initiative A	Initiative B	Initiative C	Initiative D
Criteria 1	10	5	4	3	4
Criteria 2	25	3	3	2	2
Criteria 3	35	3	1	2	2
Criteria 4	30	1	2	4	1
Weighted average	(100)	2.6	2.1	**2.7**	1.9

Figure 2.6 Weighted ranking prioritization.

The initiative with the highest weighted score is preferred first, taking into account budget considerations, etc.

An example of the weighted ranking prioritization is depicted in Figure 2.6. Four criteria have been considered for prioritization. These can include, for example, alignment with a maintenance objective, complying with a regulatory standard, improving internal efficiencies, and contributing to skill sets. Percentages (weights) have been assigned to each of the four criteria for evaluation. Each initiative is scored from 0 to 5, depending on its alignment with the criteria. Weighted averages are then calculated for each of the initiative. As per these calculations, initiative C with the highest weighted score of 2.7 is the first preference for investment. It is followed by initiatives A, B, and, lastly, D. Again, the weighting factors and the scoring models need to be approved by the PSG and are used by the PfO during portfolio finalization. However, it has to be noted that management judgment is more important than the routine application of quantitative analysis techniques.

d. **Reconcile:** The initial portfolio that gets worked out after prioritization is usually not the optimum one. This is especially true when the portfolio requires reconciliation across available funding limitations, resource constraints, etc. Many companies separate out budgets for the implementation of change initiatives, as compared to ongoing operational expenses. As per Corporate guidance, the budgeting for the change initiatives is governed by business case discussions, stage/gate review–based phased funding, accountabilities, and toll gate reviews. The timing of the execution of change initiatives is critical as their implementation lifecycle is linked to the benefits lifecycle. And as repeatedly stated, every organization has a limited appetite for change (i.e., all change initiatives cannot be implemented in parallel in an organization as it will upset their capability to handle operations). There are ongoing change initiatives and Business As Usual (BAU) activities to be taken into account while incorporating new changes. This reconciling can include balancing the following:

- Transition schedules of diverse initiatives across diverse SBUs/geographies
- Integration with the business change lifecycle of the organization
- Training/skill set enhancement requirements of various units, without introducing too many disruptions
- Pilots/rolling wave implementations, so that the "easier" geographies are tackled first—their lessons being incorporated for later implementation

Usually, a Portfolio Implementation Plan is created by the Portfolio Manager and the PfO, stating which components of the portfolio need to be executed when and which resources are needed for this execution. The implementation plan also needs to create a dashboard showing a "see through view," highlighting which initiatives are in progress, the resources assigned for the initiatives, risks, benefits to be realized from the various initiatives, and the contribution to the strategic objective (aligned with the balanced scorecard perspective, if applicable).

In order to monitor the progress of initiatives through the dashboard, the veracity of data needs to be ensured, as the data elements can come from multiple sources. It is the responsibility of the PfO to design the dashboard (especially in consultation with the PMG). The PfO obtains the data from multiple stakeholders and presents it in a useful format for decision making at various levels. When these data elements are unconnected, the organization can easily slip into a “data rich and information poor” syndrome, which has to be strictly managed.

2.12 Portfolio Funding

The funding model of the portfolio plays a critical role in implementation. Some companies take the “top-down” approach, wherein the initiative and investment decisions are taken at the central level and communicated to the respective SBUs. This approach typically works well in companies with centralized control. It enables a better centrally controlled alignment with SBU-level initiative management and progress tracking. On the other hand, the SBUs may feel that critical decisions have been taken at the Corporate level, and they do not have much “say” in initiative finalization. Common reporting templates and workgroup collaboration tools enable smoother synchronized reporting and progress tracking.

In some companies (especially in holding structures), the group company only specifies the targets for the financial perspective, and the constituent companies define strategic initiatives at the constituent level. This approach is more applicable if the SBUs are having diverse business operations. The initiatives to be taken for shared support functions (such as HR/IT, etc.) generally cater to the requirements of core functions (e.g., manufacturing, sales, and marketing), and the success of these internal initiatives is usually determined by the service-level agreements agreed upon with corresponding core functions.

Some other companies use a model in which investments crossing a particular threshold value are centrally controlled, and others are delegated to the SBU or the geography levels. In such a model, portfolio inventories and prioritization can focus on centralized investments first. The funding allocation at the central level can align with the relative strategic priorities and expected contributions of these initiatives to the overall organization.

2.13 Portfolio Optimization

Portfolio optimization consists of the selection of the appropriate mix of components that hold the “best chance” or potential to achieve the strategic objectives. This mix also depends on balancing the short-term gains with longer-term objectives, allocation across multiple types of initiatives, matching with organizational capacity, and capability bandwidths.

Some companies use the concept of Markowitz's "portfolio efficiency" frontier[9] to decide which portfolio mix is preferred. In accordance with this technique, different initiative combinations are plotted on a graph, with the X axis representing the risk/volatility of the portfolio (usually measured by its standard deviation) and the Y axis representing the expected return of the portfolio. The organization also determines a frontier, connecting the acceptable levels of risk and return. The optimum portfolio is one having the lowest risk for a given target return (or the highest return for a stated threshold of risk). Optimum portfolios coincide with the points of the frontier curve (for a stated risk threshold or for the expected return). The risk appetite of the company determines if it is willing to settle for a "low risk/low return" portfolio, or a "medium risk/medium return" or a "high risk/high return" point along the efficiency frontier. Group conferencing techniques are used in decision making on the composition of the portfolio, especially if the company is large, operating across multiple geographies with diverse product lines.

Some other companies also take the view that few of the change initiatives relating to concept development/feasibility studies can be centrally funded (or funded at the respective SBU levels) and are not included in the master portfolio. Once the decision is made to continue these initiatives into the component mainstream, they are then included in the active portfolio set and monitored. One advantage of this approach is that the portfolio does not have too many initiatives that have not been proven in concept and might not be continued. However, since all initiatives consume efforts and money, the PSG may like to set guidelines on resource allocation for such development pipeline initiatives. The set of initiatives to be pursued also depends on the overall organizational energy—how much effort and bandwidth it can give for initiating, planning, monitoring, and reviewing the change initiatives.

The budget for implementing the change initiatives can be maintained at the organizational level or at the SBU level. It needs to be taken into account that this budget is separate for maintaining operations as well as the procurement of capital assets, which could be useful across multiple initiatives of the organization.

2.14 Portfolio Implementation

After portfolio formulation comes the challenge of its implementation. Strategy execution remains as a top challenge in the CEO's agenda. Rather, one could say that the very core of the CEO's functions is to design the organization's strategy and facilitate its execution through the portfolio. Diverse factors underpin this challenge. These include very low understanding of strategy by the workforce, middle managers' job descriptions not being aligned with the strategy, misalignment of budgets with strategy achievement, and top managers not devoting adequate time to monitor strategic achievements.

Portfolio implementation requires much more attention and effort as compared to planning. Herein lies a commonly noted fallacy. Most of the Senior Managers leave the implementation of the portfolio to middle-level managers, assuming things will work out by themselves. Execution involves more efforts than planning, and coupled with the constraint of implementation resources being shared with the BAU, the prioritization between portfolio implementation and sustaining business operations is a challenge. In addition, whereas the focus of the top managers may be on strategy execution, middle-level, and workforce employees usually have day-to-day operational challenges to resolve. It is indeed a very desirable characteristic of top management not to "lose the sight of woods for the trees." Many effective managers are endowed with "helicopter vision" and are able to see the big picture along with the ability to deep-dive into details, where necessary.

During portfolio implementation, a critical aspect to be considered is that the initiatives in the portfolio should continue, to remain aligned with the strategy. For instance, if a company chooses a cost competitiveness route to garner more market share, the initiatives it takes need to be aligned with this strategic objective (as compared to, for example, market diversification or focusing on niche markets). In case the strategy changes, it becomes the responsibility of the Portfolio Manager/PfO to retune the portfolio and obtain approval from the PSG. The progress in achieving the strategy is monitored by the PMG, which can also give reverse feedback to the PSG.

This point is more applicable for the programs under the portfolio, as generally large programs run for multiple years and benefits take considerable time to be realized.

For programs, the outcomes and the early benefits are assessed and the portfolio progress reports are updated. The portfolio implementation plan (prepared by the Portfolio Manager and approved by the PMG) also includes the governance arrangements and the stakeholder engagement/Risk Management Plan to be invoked during implementation of the portfolio. The COE attached to the PfO maintains past lessons and best practices, which need to be referred to by the Portfolio Component Managers before launching their components. The governance arrangements also include invoking the criteria regarding when to initiate a component, when to terminate it, what type of toll gates will be applied and how the viability of an initiative will be assessed and reassessed on an ongoing basis. Essentially, governance provides answers to the questions, "Are the right change initiatives being undertaken, are these being executed correctly, and are the benefits being realized?," on an ongoing basis.

As the components under the portfolio are being implemented, it is quite likely that critical risks and issues will need to be addressed by the Portfolio Manager. Whereas the portfolio governance plan (which can be part of the portfolio implementation plan) gives guidance on the management of key risks and issues, the Portfolio Office is expected to assist the Portfolio Manager by assessing dependencies/other factors and evaluating the impact on the strategic objectives of the organization. On the basis of the progress reports, the Portfolio Manager can reallocate resources, expedite or slow down a portfolio component, or even prematurely terminate a component.

Performance dashboards are used by many organizations to view and aggregate the progress data for producing views and reports suitable to various levels of management. The Portfolio Office can maintain and update the dashboard.

During portfolio implementation, it may be necessary to communicate the progress to various stakeholders. For government or publicly listed/external stakeholder–funded initiatives, it will be necessary to send statutory reports, as well. Typically, it becomes the responsibility of the Portfolio Sponsor to communicate the status of the portfolio (which could represent the business value of the company, as well), forecasts, and any major issues and risks to the Senior Management/PSG. As a part of the portfolio implementation plan, the portfolio communication plan contains guidelines on what to communicate, to which stakeholders, through what types of media, etc. The PfO can include a service to facilitate holding communication events—for example, town hall meetings and investor conference calls. These communication events become powerful facilitators for stakeholder engagement as well.

Likewise, risk management is crucial at the portfolio level. For strategic risks, the Portfolio Sponsor becomes the risk owner for addressing the risk with inputs from Senior Managers. Some of the risk events that can potentially undermine the portfolio include:

- Expected changes in government policies (such as nationalization of banks and devaluation of currency)
- Impending changes in legal regulations (such as the Sarbanes-Oxley reporting requirements when they were first announced)
- Envisaged natural or man-made events (such as impending tornadoes or terrorist attacks, etc.)

The portfolio risk management plan (which is a component of the portfolio implementation plan) gives guidance to the Portfolio Manager on the process of risk management. The Portfolio Risk Register populates the risks on an ongoing basis, and these are addressed by the Portfolio Manager along with the respective risk owners. As in project management, techniques such as sensitivity analysis, "Monte-Carlo" analysis, and others, can be used to assess the impact of the risks under different assumptions. The risk response plan at the strategic level is critically dependent on the risk appetite of the company, as many of these risks have an overarching influence across multiple change initiatives in the portfolio.

Programs and independent projects are launched as a part of the portfolio components, and the resources are allocated for these initiatives. The portfolio governance plan (as a part of the portfolio implementation plan) provides the checks and controls that need to be taken into account while launching new initiatives, how to oversee and close them, and integrate outcomes back to the BAU.

The portfolio governance plan provides guidelines on when to revise the portfolio. The business change lifecycle of the organization also determines how frequently the portfolio needs to be updated. The portfolio governance plan incorporates mechanisms to provide assurance to the PSG that proper procedures are being followed for portfolio component initiation, funding of the initiatives, and their progress tracking systems/gated reviews to ensure that the components are aligned with the set standards.

The progress reports from the programs/projects provide input to the Portfolio Manager in tracking the portfolio progress through dashboards and other reports sent to the PMG. The PfO can assist the Portfolio Manager by collecting data and putting together the reports.

The success of the portfolio itself can be measured by the business value it provides to the organization. Additionally, facilitating the right initiatives to be taken up and monitoring them to ensure they are implemented correctly are other benefits of robust portfolio management.

The Portfolio Office can design metrics to measure how successful the portfolio management process is by showing the trends in spending across multiple portfolio categories, how many initiatives have not been taken up or proceeded with because of viability factors, improved resource utilization rates, etc. An enhanced reputation for the organizational governance, especially in the eyes of regulators, auditors, funding agencies, and shareholders is a valuable benefit of robust portfolio management.

However, in order to have consistent portfolio success, it is essential to have well-defined procedures for program and project management embedded in the organizational culture.

We will discuss how to manage programs and projects in subsequent chapters.

References

1. Henderson, B. (1970). "The Product Portfolio." Retrieved from https://www.bcgperspectives.com/content/Classics/strategy_the_product_portfolio/.
2. Aaker, D. A. (2001). *Strategic Market Management*. New York: John Wiley & Sons. 108 pp.
3. *Porter*, M. E. (1980). *Competitive Strategy: Techniques for Analyzing Industries and Competitors*. New York: Free Press.
4. Ibid.
5. Porter, M. E. (1985). *The Competitive Advantage: Creating and Sustaining Superior Performance*. New York: Free Press.
6. Ibid.
7. Kaplan, R. S. and Norton, D. P. (1996). *The Balanced Scorecard: Translating Strategy into Action*. Harvard Business Review Press.
8. Saaty, T. L. "Relative Measurement and Its Generalization in Decision Making: Why Pairwise Comparisons are Central in Mathematics for the Measurement

of Intangible Factors—The Analytic Hierarchy/Network Process." *RACSAM* Vol. 102, No. 2 (2008): 251–318.
9. Markowitz, H. "Portfolio Selection." *The Journal of Finance* Vol. 7, No. 1 (Mar., 1952): 77–91.

Chapter 3

The Core of Program Management—Benefits Management

3.1 Program Management—The Context of Benefits Management

Portfolios are executed through the implementation of programs, as well as individual projects linked to the portfolio. In this chapter, we focus on how to plan and implement programs.

Programs consist of a collection of interdependent projects to produce outcomes and realize the required benefits. Outcomes and benefits result from changes in business operations.

In complex programs, there could be a considerable time lag between the transition of project outputs and the realization of benefits. This is especially applicable for industries with "low velocity," such as infrastructure creation (e.g., commissioning a power plant).

Governments also execute programs that benefit the social sector—for example, capacity building, enhancement of healthcare, and improvement in literacy rates. These could be noncommercial but vital to enable the countries to become more competitive and increase the standards of living for their respective populations.

The aim for every program is to understand which benefits need to be realized to achieve the strategic objectives of the organization. This could also be applicable for the not-for-profit and governmental sector—with social objectives in place. In Chapter 2, we discussed on how organizations define their portfolios (which include the change initiatives to be taken up as programs and independent projects) to achieve their strategic objectives.

Benefits management forms a critical linking pin between portfolio management and program management. Typically, benefits are identified at the portfolio level, whereas program management can determine additional benefits that can be obtained from implementing change. The extent of investments allocated by an organization for program management in a planning horizon depends on multiple factors.

These factors can include:

- The maturity of the initiative appraisal and selection process
- Track record of success of past initiatives (especially programs)
- Organizational governance processes adopted
- Enterprise culture and its risk appetite

There are a couple of techniques used for benefits identification during portfolio management and program management. These techniques can include the creation of the following:

- Benefits Logic Map
- Results Chain
- Benefits Map

We present only a brief overview of the "Benefits Map," as other techniques are somewhat similar to it.

3.2 Benefits Map

The benefits map is a unidirectional dependency map, showing the project outputs noted towards the left-hand side, leading to the intermediate benefits, which in turn link to the end benefits envisaged. These end benefits, in turn, lead up to the strategic objectives in the extreme right in the benefits map. An illustration of the benefits map is shown in Figure 3.1.

In this illustration, five projects are shown leading to four intermediate benefits/outcomes. It is also noted that the first three project outputs combine to create a capability (controlled by the Program Manager). The capability is transitioned to Business as Usual (BAU) to realize the outcomes/intermediate benefits. This further leads to the "More revenues" end benefit, facilitating achievement of the strategic objective of becoming the market leader. The last two projects lead to two further outcomes, resulting in the realization of "More profits" being the end benefit, which also links up to achieving the same strategic objective of becoming a market leader.

Typically, the benefits map is initiated with strategic objectives in mind (which can be traced back from the portfolio) and deliberating which benefits need to be obtained to achieve the strategic objectives. The Portfolio Office (PfO) referred to in Chapter 2 can facilitate decisions as to which outcomes/intermediate benefits can lead to the attainment of the stated benefits. The balanced scorecard techniques can be useful here, as most of the intermediate benefits can be tracked to learning and growth, internal business processes, and customer perspectives.

3.3 Multiple Ways a Program Can Come About in an Organization

There are multiple ways in which a program can be conceptualized. Some of the programs are driven "top-down" from senior management. This scenario is more likely when there is a leadership change at the top and can also be due to mergers and acquisitions. In the case of governmental programs, a change in political leadership can trigger off such programs.

In established organizations, there could be ongoing change initiatives addressing the same set of objectives or benefits. The organization may deem it useful to group them under a single program umbrella for synergy and better coordination. In this case, the benefits map may have some ongoing projects already included before the program officially starts.

Which initiatives are to be taken up to achieve the outcomes and realize the benefits fall under the realm of program management. The

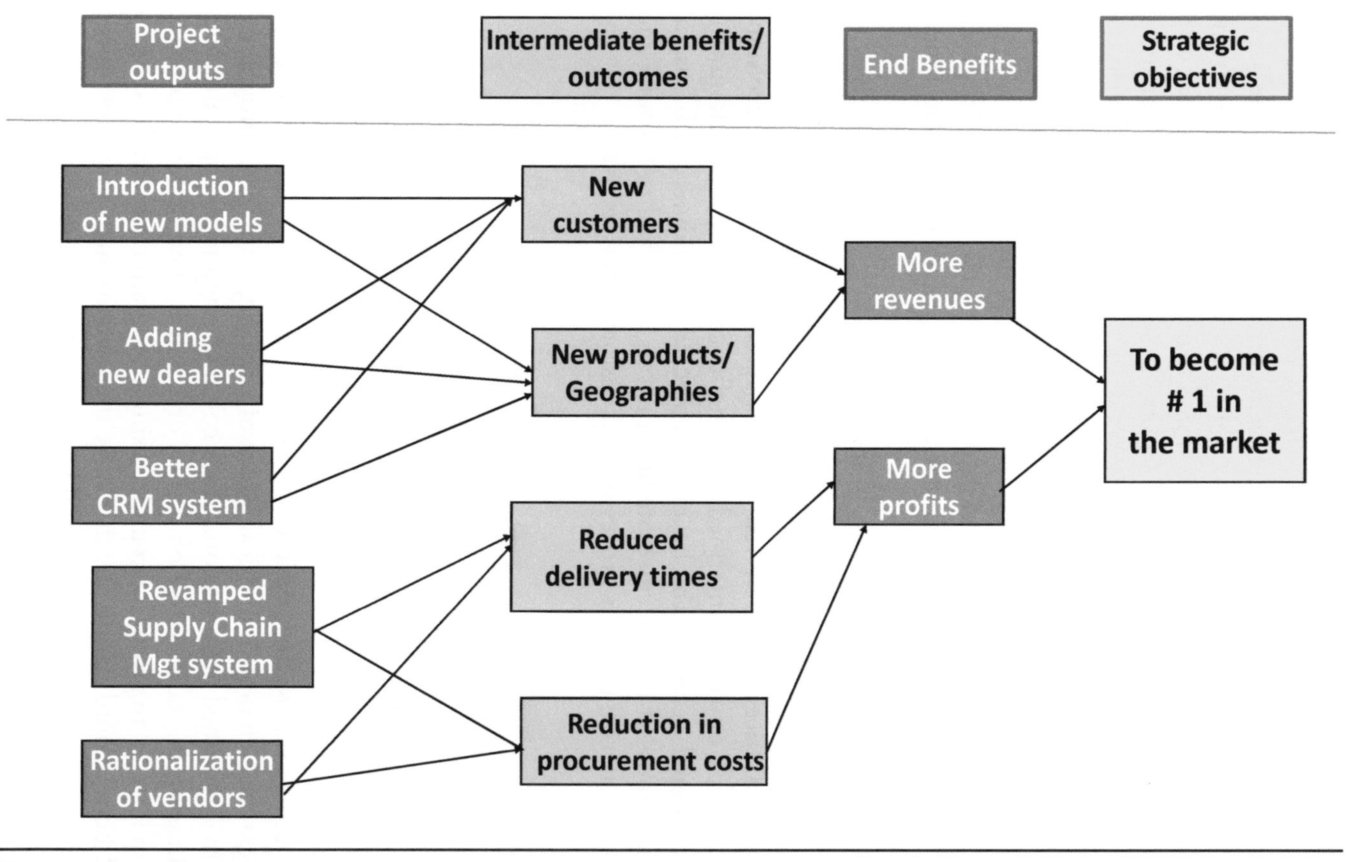

Figure 3.1 Benefits Map.

trigger for most of the programs is a program mandate, produced by senior management.

3.4 Program Mandate

The Program Mandate informs why the change initiative could be considered to be run as a program. It can include details on the triggers for the program to be taken up, strategic objectives supported, benefits and outcomes envisaged, an overview of the timelines and budgets within which the program needs to be implemented, and the overall governance arrangements within which the program will be guided. In many cases, the program mandate contains a "high-level business case" for initial consideration and elaboration.

The mandate will usually also include reference to the Program Sponsor (which, in many companies, is also called the Program Director), who becomes accountable for the success of the program. The Program Sponsor belongs to the Senior Management Group, which could be part of the Portfolio Steering Group (PSG) as well as the Portfolio Progress Monitoring Group (PMG). The program mandate is a precursor document to full-fledged program approval, which occurs when the Program Charter is subsequently prepared and approved.

The Portfolio Office identifies the program during its portfolio definition, which can then lead to the program mandate. The program governance, in turn, is intricately linked to the reporting and oversight requirements of the PMG.

Preparation of the program mandate provides an early indication to the performing organization to consider if it needs to be investing in it all by undertaking feasibility or exploratory studies. By getting more information (which can relate to market demand, emerging technologies, competitor stances, etc.), the organization has more data-points while the program charter is getting prepared for informed decision making.

The program mandate is discussed further in executive meetings regarding its likely viability and what constitutes the acceptance criteria for the success of the program. This is a positive step, as consideration of the program mandate gives an initial idea of the viability of the program. It is quite likely that the unviable or "pet" ideas get discarded during the ratification step of the program mandate. Typically, during the program mandate confirmation, the Program Sponsor is formally involved, but the Program Manager may not be.

3.5 Program Governance Board

As a part of the program mandate ratification, top management also constitutes the Program Governance Board. This board usually consists of the Program Sponsor (as the head of the board), senior Functional Managers (who would be impacted by the program outcomes and provide resources), representatives of Corporate shared services, etc. The Program Manager usually takes on the role of the convener of the Governance Board meetings.

The Program Governance Board is the highest decision-making body within the program lifecycle. Major responsibilities of the Governance Board include the following:

- Approving the initiation, transition, and closure of various components
- Approval of major deliverables across major phases of the program
- Providing resources and funding for the program
- Dealing with major escalations (including risks and issues) from the Program Manager
- Communicating the status of the program to top management and to major external stakeholders (such as funding organizations, shareholders, etc., as applicable).

Usually the Program Sponsor takes the key role in these high-level communication events.

3.6 Program Lifecycle—Phases

The program lifecycle can be divided into four main phases:

1. *Program initiation:* The program mandate is the input, and the program charter is the output.
2. *Program definition:* The program definition takes the program charter as the input, and the Program Management Plan is the exit document.

3. *Program execution:* This is where the actual program is implemented in multiple iterations. There are two sub-phases in this phase, which interface with each other. These include:
   a. *Component initiation, oversight, and integration*: This initiates the components relating to each iteration; and
   b. *Benefits realization*: This focuses on transition management, outcomes, and benefits management.
4. *Program closure:* This phase closes the program and transfers the overall set of benefits to the BAU.

The program closure phase can also occur because of an abnormal termination of the program due to the withdrawal of funding, a change in business strategy or the rationale for the program, etc.

A summary diagram outlining the phases is presented in Figure 3.2. The details of work involved in each of the phases are given below.

3.7 Program Initiation Phase

After the program mandate is approved, the next major deliverable, which is produced in the program lifecycle, is the Program Charter. The program charter is a pivotal document that "kick-starts" the program and, in a way, is the first formal document produced within the program lifecycle. This charter can contain the following information:

– Name of the program
– Names of Program Manager/Program Sponsor
– Strategic alignment of the program within the organizational context (thus linking it with the portfolio)
– Program outline vision statement
– Program outline business case
– An overall indication of the scope, schedule, and budget for the program
– Benefits expected from the program
– Constraints/assumptions/major risks and an indication of how they are proposed to be addressed
– Recommended program governance structure
– Extent of stakeholder support/stakeholder considerations

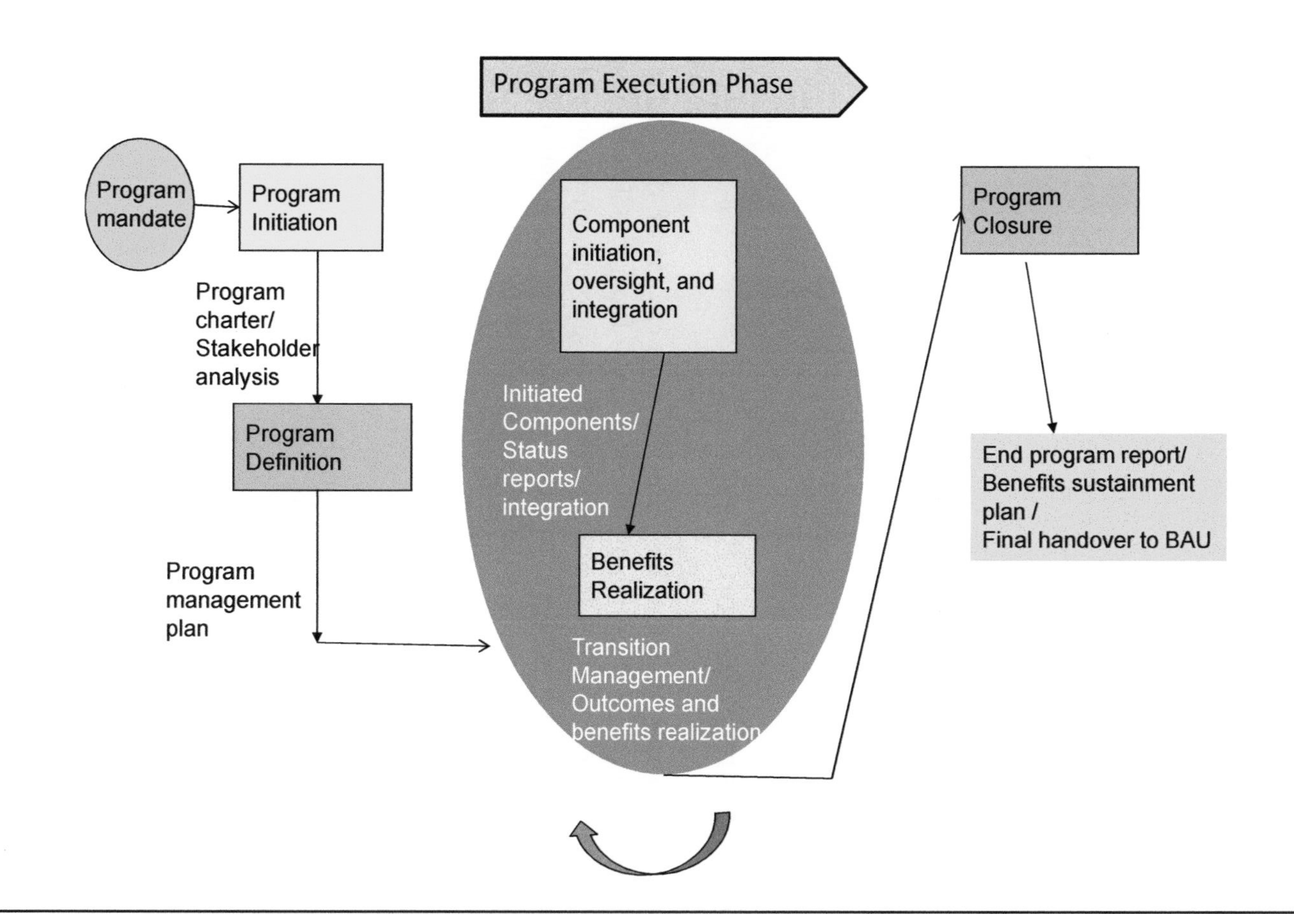

Figure 3.2 Program lifecycle—an overall view.

- Initial list of component projects and their business cases (if applicable)
- Program success criteria

The program charter can be prepared by the prospective Program Manager but would need the approval of the Program Sponsor and the Portfolio Steering Group (PSG).

The following factors would need to be considered, among others, before the program charter is approved:

- Does the program address benefits that are required to be realized to achieve the portfolio strategic objectives?
- Are there existing programs seeking to obtain the same benefits? If so, what is the value added if this new program is launched?
- Does the organization have the capacity and capability required to resource the program?
- Is the cumulative risk by taking up the program within the overall risk appetite? (Organizations executing too many disparate programs tend to scatter their efforts, losing focus.)
- What has been the success of previous change initiatives? Is the organization ready for yet another large change initiative?
- From the outline business case produced, does the program look viable?

It is quite possible that the program mandate is produced, but based on the program charter, the PSG decided not to go ahead with the program as it looked unviable or did not fulfill some of the criteria listed above. This is significant, as this decision during the early stage of the program lifecycle helps to avoid the preparation of time-consuming and detailed cost analysis, Work Breakdown Structure (WBS), etc., when the fundamental idea of the program was not viable—for instance, the benefits may not be forthcoming within the schedules expected from the program. In such cases, the program charter is "archived," stating the reasons why it was not approved by the PSG. It is quite likely that some programs revive when conditions change. In this context, previous work invested in charter preparation would be valuable for updates later and to obtain the required approval.

The program charter may also contain a reference to the "outline vision" statement, specifying the end goal of the program, which indicates the future state of the impacted organization after the execution of this change initiative. The vision statement is refined further during the program definition phase, incorporating the inputs received from multiple stakeholders.

The program business case balances the benefits to be accrued from the program and the likely costs thereof. It needs to be noted that program business case costs cover multiple aspects, including the program component-related costs, benefits realization and measurement costs, and the costs concerning the program management itself. The program business case is updated during the program execution phase, "outlives" the program, and is handed over to the BAU during program closure.

Approval of the program charter is typically the responsibility of the Program Governance Board. Once the program charter is ratified, the Program Manager takes over much of the planning work.

3.8 Program Stakeholder Engagement

The initial work undertaken by the Program Manager is to ascertain the stakeholder expectations from the program and also develop an interest-influence matrix. Such a map classifies the stakeholders according to their current interest and influence and develops engagement strategies with different types of stakeholders. The map is dynamically updated during multiple phases (especially during the sub-phases of the program execution phase). The Program Manager would normally consult the Senior Managers (especially the Program Sponsor) while updating the interest-influence matrix and deciding which steps to take to obtain buy-in from senior stakeholders.

It would be naïve to assume that all the stakeholders would want the programs to succeed, especially those stakeholders who are negatively impacted and would like the program to fail. Such "negative stakeholders" need to be especially watched by the Program Manager.

A representative stakeholder interest-influence matrix is shown in Figure 3.3. As stated in this matrix, the Program Manager needs to monitor the stakeholders in the "low-interest/low-influence" quadrant, as some of them may turn out to be "negative" stakeholders. Stakeholders who have high interest in the outcomes of the program but little influence over its direction, etc., need to be "kept informed," as some of them can turn into internal coaches for the program. Stakeholders with low interest

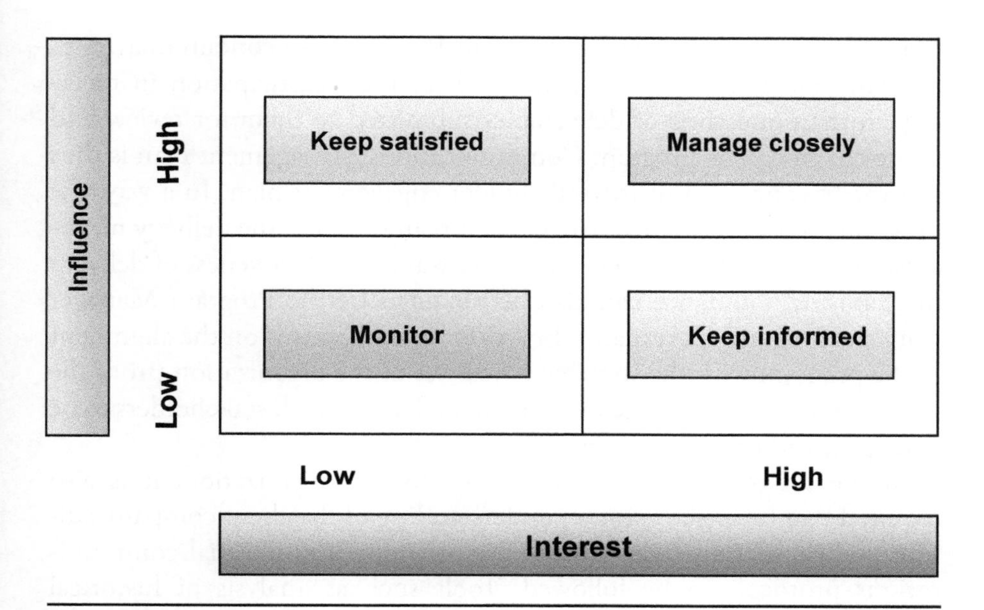

Figure 3.3 Interest-influence matrix.

but having a high influence in the program outcome need to be kept satisfied—through compliance reports and understanding of their objections, etc. The most important quadrant for the Program Manager is those stakeholders who fall into the "high interest and high influence category." It is imperative that all the positive stakeholders are found in this quadrant. A negative stakeholder falling into this quadrant poses a high risk for the success of the program. In such an event, the Program Manager needs to enlist the assistance of the Sponsor, if needed, to address the objections of that stakeholder and to actively listen to his or her concerns and address them, if possible.

Different communication events can be planned by the Program Manager for diverse types of stakeholders, with varying communication channels. For instance, the stakeholders in the low-interest/low-influence quadrant may be receiving routine progress reports, but the stakeholders in the high-interest/high-influence quadrant would require regular, intensive, and bidirectional communication. And the stakeholders with high interest and low influence may receive customized communications showcasing the program's progress, which can be further propagated by them to the concerned functional stakeholders to enable getting a better buy-in for program success. The Stakeholder Register with the Stakeholder Engagement Plan would be prepared and updated by the Program

Manager. The Stakeholder Engagement Plan can also contain metrics to measure stakeholder involvement (such as their participation in meetings, turnaround time of deliverables submitted to them for review and approval, etc.). The program Communications Management Plan is thus intricately connected to the stakeholder engagement plan. In a way, the program communications management plan becomes the delivery mechanism to engage stakeholders, and it portrays the effectiveness of delivery mechanisms. Guidance can also be obtained by the Program Manager from the Program Governance Board/Program Sponsor on the alignment of the program with the strategic objectives of the organization (from the portfolio perspective), outcomes and benefits expected, stakeholders to be managed, and other key information.

If the Program Manager is external to the organization, it is also required that he or she gains an understanding of the client company culture, stakeholder attitudes toward the program, and internal communications protocols to be followed. Tools such as analysis of historical information, interviews, focus group discussions, etc., are useful for the Program Manager to uncover stakeholder expectations.

It may also be useful to classify the stakeholders into multiple groups, such as users/beneficiaries from the program, suppliers, governance, and external influencers (e.g., external funding organizations), for a better understanding of stakeholder interests. This is further elaborated in the discussion in Chapter 6 on stakeholder engagement.

The work done so far can be grouped under the "program initiation" phase.

3.9 Program Definition Phase

During the program definition phase, the detailed program management plan is prepared by the Program Manager. The program management plan is a consolidated plan, covering benefits realization, scope, schedule, financial management, quality, resource, risk, issue management, etc. We cover some of these plans here in greater detail:

A. Benefits Realization Plan:

This plan shows when the benefits for the program are expected to be realized, along with the likely dates of outcomes. This plan outlives the program management plan, and once the program gets closed, it is passed on to the BAU for maintenance. Although the Program Manager is the document owner for this plan, significant inputs need to come from the Functional Managers (who could be the owners for the benefits). The Benefits Realization Plan is produced during the program definition phase, updated during the program execution phase, and handed over to the BAU during the program closure phase. This plan can also contain the Transition Plan, stating when the transitions need to occur during the program, the roles of multiple stakeholders during transition management, and aspects that need to be kept during transition. If the components of the program are delayed, it is likely that the program benefits are also affected. This synchronization needs to be kept in perspective by the Program Manager and the Functional Managers alike.

B. Program Scope Statement:

Whereas the program charter gives an outline of a scope statement, it is the responsibility of the Program Manager to prepare the scope statement in detail. Although tools such as interviewing, etc., would be more useful at the project level for scope management, at the program level, the judgment of the Program Manager on how to achieve the program objectives through decomposition into components is more important. It is also quite likely that the Program Manager may not be able to completely chart out which components the program needs to execute during the commencement of the program. Components can be added (and even terminated prematurely) by the Program Manager, depending on how the program is progressing.

The Program Manager can create an initial Program Work Breakdown Structure (PgWBS) during the program definition phase to depict the major components that will be executed. While doing so, the Program Manager will confine the decomposition up to the first two levels of the Project Work Breakdown Structure (PjWBS). The detailed PjWBS will be prepared by the corresponding Project Manager as a part of the project management lifecycle.

It has to be noted that the PgWBS (and any WBS for that matter) consists of only deliverables (and not activities, which will be detailed during schedule development). Some of the companies call the WBS the Product Breakdown Structure (PBS), but we will use the term WBS consistently. The advantage of first producing the WBS before the detailed activity development is that WBS affords clarity on which deliverables to be produced first (rather than mixing with it with the question, "How do we do that?"). At the program level (and at the project level, as well), the work flow up to the WBS determination is usually sequential. Thereafter, schedule and cost estimation, risk and quality management, etc., occur concurrently and iteratively. This is why WBS becomes the pivotal document in the project (and in the program) environment. The Program Scope Baseline can include the overall program scope statement and the PgWBS. Some companies also produce a Scope Management Plan detailing the guidelines on how to prepare and update the scope statement and the WBS.

C. Program Schedule Development:

After the development of the PgWBS, program master schedule development is invoked. For each of the PgWBS "node" elements, activities are determined and sequenced. Some companies develop an overall summary "Program Roadmap," which identifies major milestones in program and benefits delivery. As was noted earlier, the Program Manager treats each component (including projects and other related work) as a "black box" and does not get involved with their detailed task execution. The program master schedule thus identifies dependencies across major components that the Program Manager needs to especially focus his or her attention on. The program master schedule may also encompass the concept of the critical path, representing the "longest path" in the master schedule, passing through the "critical" components that require more attention from the Program Manager from a schedule perspective. The program master schedule finalization is done iteratively with that of the program cost baseline, balancing the availability of resources and any deadlines. It is also important to keep not only the likely end date of the program in perspective but also any intermediate milestone deadlines and constraints. It may also be noted that programs are not typically time constrained, as compared to projects. From the program environment, success factors are measured more by the realization of outcomes and benefits.

It should also be noted that there is an element of back and forth interactions between the Program Manager and the concerned Component Managers during schedule finalization. The Program Manager can work backwards, considering the overall envisaged end date for the program and assigning a corresponding likely end date for the components.

This information on the likely component end dates is analyzed by the Component Managers (along with component scope and budgets) and gives reverse feedback to the Program Manager. The Program Manager reassesses the program schedule, with the inputs given by the Component Managers and the dependencies across the components, to finalize a viable schedule. In large programs, this interaction can be complex, and the Program Management Office (PgMO) needs to support the Program Manager in its preparation. The detailed roles of PgMO (in the context of the programs launched as a part of the portfolio, etc.) are discussed in Chapter 8. Large companies also use Enterprise Program Management (EPM) tools during finalization and updates of the PgWBS and the program schedule.

D. Program Financial Management Plan Finalization:

The Program Sponsor is expected to secure the funding for the program. This can be involved if the money needs to be obtained from diverse sources, such as through public funding, private equity funding, etc. In this case, the weighted average cost of capital comes into consideration regarding how much funding needs to be obtained from which source. Whereas the lowest cost of capital may be alluring, sometimes the program funding needs to be planned considering the requirements of initial funding to fast track benefits realization. A balanced approach may need to be taken in such situations.

The financial framework for the program may need to consider how much money the Program Manager/Sponsor needs to mobilize from which sources and the likely fund flows. This analysis (covering likely fundings and expenses) can alter the program master schedule and the benefits realization plan. The program cost performance baseline (as in an S-curve for the project) can then be baselined after synchronizing with resource availabilities and external constraints.

E. Program Quality Management Plan:

The Program Quality Management Plan defines the minimum standards for quality to be applied to its components. Usually, this plan is derived to be in alignment with Corporate quality standards and the regulatory frameworks the company is expected to follow. The program quality management plan focuses on process compliance rather than product correctness (as in project-level quality). Thus, the program quality management plan may test to ascertain if the Program Manager and other concerned stakeholders are adhering to stated quality standards and plans (e.g., if the stakeholder engagement plan is executed appropriately). At the program level, the quality management describes the performance of assurance reviews, health checks, etc. The PgMO can facilitate quality assurance. However, the PgMO personnel performing quality assurance need to be distinct from the team that is engaged in data support functions, as stated in the detailed functions of the PgMO covered in Chapter 8.

F. Program Risk Management Plan:

At the program level, the risk management plan contains guidelines on how to identify, prioritize, respond to, and monitor risks. Again, this risk management plan synchronizes with the Corporate standards for risk management. A Program Risk Register is used to capture program-related risks. Such risks can broadly have the following attributes:

- Risks unique to the program, such as risks expected during transition management
- Cross-cutting risks from the components. Some risks can impact multiple components, in which case it is appropriate to address the risks at the program level. The risk management plan at the program level (and the individual project risk management plans) contains the thresholds for escalation between the components and the program.
- A significant risk arising out of a component, for which the Component Manager does not have the authority to address. In such cases, the risk is escalated to the Program Manager, who does the impact analysis. This analysis may require the related Component Managers to be consulted by the Program Manager.

In addition, at the program level, individual component-level risks aggregate (both from threats and opportunities). Thus the same event that can be a threat to one component can become an opportunity to another component (as is likely in funding reallocations across components that are due to change in strategy, etc.).

As at the project level, the program-related risks could be identified by any concerned stakeholder. These can include the Program Sponsor, the

PgMO, the Component Managers, vendors, or the support functions. Any role that identifies a risk becomes a risk author, and once he or she identifies a risk, the risk is recorded in the program risk register. The Program Manager, along with the risk author and other Subject Matter Experts (SMEs), can prioritize the risks. Usually, the risks are prioritized across three dimensions—probability, impact, and the proximity (time point when the risk is expected to occur first). Once the risks are prioritized, they are classified in a program probability-impact matrix (PI matrix). A sample PI matrix is depicted in Figure 3.4.

In this figure, risks with a high probability and severe impact are classified in quadrant I. Quadrant II contains risks that have either medium probability and severe impact or high probability and medium impact. (The guidelines for classification of the risks are stated in the program risk management plan.) Some of the grids may contain more than one risk, and some grids may not have any entries at all. It may be noted that at the project level as well, a similar PI matrix is applicable. In addition, the risk register is continuously updated during the program lifecycle.

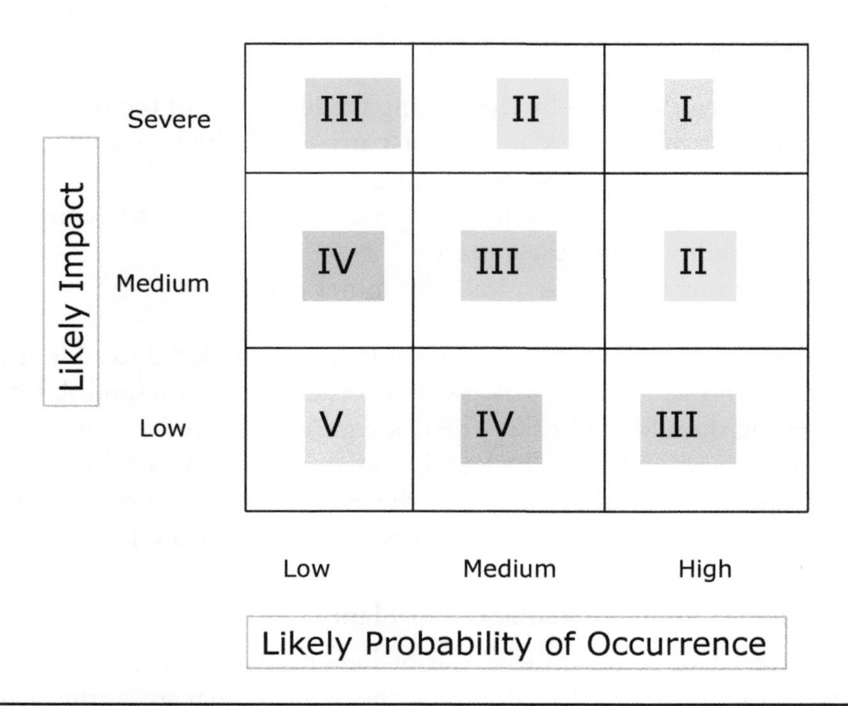

Figure 3.4 Probability-impact matrix.

The program risk register can include the following details for each risk:

- The risk identifier
- Risk cause/description of the risk event/risk effect/early warning indicator of the risk
- Probability/impact/proximity (including its states—when it was initially identified—and the current state—as it gets updated on an ongoing basis)
- Expected monetary value of the risk (which in many cases, defined as the probability multiplied by the impact—including its updated value)
- Details of the risk author/risk owner
- Risk response applied (including description of the secondary risks, if applicable)

It should be noted that the risk register is initially prepared during the program definition phase and is continuously updated throughout the life of the program.

Once the PI matrix is prepared, the risk responses for risks in various quadrants can be formulated. These risk responses can include:

- Avoiding the risk—by eliminating the root cause of the risk
- Transferring the risk to a third party—including subcontracting/entering into an outsourcing agreement, etc.
- Mitigating the risk—through proactive responses to reduce the probability and/or the impact
- Accepting the risk—typically for minor risks.

As noted, for each risk, a risk owner (who is best placed to address the risk) is identified, and the risk owner is vested with the responsibility of addressing the allocated risks. The risk register needs to be continuously monitored and updated. The PgMO can assist the Program Manager in these updates. It is most likely that the overall risk exposure of the program keeps changing as it moves across the program lifecycle.

G. Program Resource Management Plan:

Resources at the program level can include human resources, computer servers, rooms, and the like. Typically, the resources that are cutting across the projects/components need to be managed at the program level. The company Resource Management Group (RMG) or a similarly named group can maintain a central repository of all the resources, their current availabilities, skill sets, etc. At the program level, this work can also be facilitated by the PgMO.

One of the critical requirements of a program resource management plan is to undertake long-term capacity planning requirements for the program and find out how to acquire them. These resources can be acquired from functional departments, contracted resources, and the like. The financial implications for each of these could be different. Since many of the programs consume extensive resources, the resource management plan needs to interface with the Program Financial Management Plan. At the program level, some unique regulatory issues can arise, such as the transfer of technology, buy-back arrangements, employment of local workers, etc., which can also constrain the Program Manager.

A critical deliverable produced during the program definition phase is the Target Operating Model of the BAU impacted by the program. This target operating model can cover the following aspects:

- Redefined processes (in the BAU) to realize the outcomes and benefits
- Redefined BAU organization structure needed to achieve the program goals
- Technology BAU interfaces needed
- Revised BAU information flows and dashboards, which are needed to measure the outcomes and monitor the benefits

This target operating model can be produced for the current organization structure (on an "as-is" basis) and the "to-be" target. The "space" between these two requires component initiatives to be completed to bridge the gap (and the Component List gets built up this way).

As an illustration, an IT service provider was implementing an Enterprise Resource Planning (ERP) package in the business functions, which was run as a program lifecycle. This program assessed the current status on an "as-is" basis and mapped what needed to be done to obtain the benefits regarding the following:

- Revised processes, including for supply chain management
- Redesigned IT organization structure—for ERP execution and data administration

- Redeployed technology, including installation of required licenses for the ERP, acquiring a new computer server, and configuring network arrangements for capturing data from remote locations for analysis
- Redesigned information flows—on sales volume, daily revenues by various Stock Keeping Units (SKUs), etc.

The ERP deployment itself was run as a system integration program in accordance with the above-mentioned elements. The "as-is" to "to-be" gap was bridged through execution in multiple iterations focusing on early benefits first to give credibility for the success of the program.

The program management plan can also include the Program Governance Plan, stating how the program is expected to be governed. The governance plan can include the following:

- Criteria by which the portfolio management will oversee the program
- Program assurance, milestone reviews, and assurance arrangements and timelines
- Escalation routes between the Program Manager and the Portfolio Manager
- How the program will govern its components (including component initiation, component progress monitoring, and component closure aspects)
- Interfacing and escalation arrangements between the components and the program

The Program Governance Board especially needs to approve the program governance plan as well as the overall program management plan (which is finalized during the program definition phase).

As we noted earlier, the list of components to be executed can depend on the target operating model (current and future operating models, gap analysis, and finalization of the components).

The program management plan needs to be approved by the Program Governance Board before the program moves on to the program execution phase. This is a critical tollgate, as once the program execution phase is approved, resource commitments become higher for the program.

The approval of the program management plan by the Governance Board marks the culmination of the program definition phase.

3.10 Program Execution Phase

The program is executed during the program execution phase. Different frameworks give varying names for the program lifecycle. We will adopt the nomenclature that the program execution phase consists of two sub-phases—the component initiation, oversight, and integration sub-phase and the benefits realization sub-phase. In reality, both these sub-phases will be running in parallel, with the first sub-phase being predominantly addressed by the Program Manager, and the second sub-phase being addressed by the concerned Functional Managers. It should also be noted that the program execution phase can iterate multiple times as needed, as noted earlier, to obtain benefits incrementally. The program business case is kept up to date, and its continued viability is assessed during these multiple iterations.

During the component initiation, oversight and integration sub-phase, components are authorized and launched by the Program Manager. The information passed on by the Program Manager to the Component Manager during component launch includes the following:

- Component scope, schedule, and costs as agreed upon after inter se consultations, as applicable
- Resource allocations
- Critical stakeholder considerations to be kept by the Component Manager related to the component
- Constraints and assumptions under which the component needs to deliver
- Major risks and issues to be addressed
- Governance/progress reporting arrangements, which will be applicable to the component (including quality assurance/reviews and acceptance criteria for the component outputs)

It can also include the quality and configuration management guidelines applicable for the component and acceptance criteria for the component deliverables:

- Component-level tolerances that are set and progress reporting requirements

- Dependencies of the specific component with other components (which could also include an extract of the benefits map concerning the component)

Much of this information can be captured as a part of the Project Charter given to the Project Managers during the commencement of their projects.

The program roadmap contains an initial list of components to be executed. An initial business case assessment could be carried out including these components in the program. However, when the components are authorized, further refinements can be carried out to finalize the project charter and approve it. (This project charter approval would usually come as a part of project lifecycle.)

During the component initiation, oversight, and integration subphase, the Program Manager obtains the component performance reports from the components. The format, periodicity, and contents of such reports would already have been communicated to the Component Managers during the launching of the components.

It is quite likely that multiple components are executed in parallel during iterations. It thus becomes the responsibility of the Program Manager to synergize their progress, reallocate the resources, and manage the risks and issues arising during component execution. The Program Manager can also provide guidance to the Component Managers as needed. In case any issues and risks are escalated by the Component Managers, the Program Manager can do an impact analysis on the program and the other components, have consultations with concerned stakeholders, and, in turn, escalate to the Program Governance Board, as needed.

During component progress tracking, it is the responsibility of the Program Manager to see to it that the components do not get misaligned with program objectives, and that the program itself remains in alignment with Corporate strategy. In case of changes in corporate strategy (such as in the Corporate vision or redesigned benefits, etc.), it is the responsibility of the Program Manager to redesign the set of components to be launched or reconfigure the ongoing initiatives appropriately.

Large programs usually see the production of multiple artifacts. So the Program Manager needs to ensure that appropriate configuration management systems are in place to enforce version control.

The PgMO can support the Program Manager in ensuring the master copy maintenance and protection of program assets.

Details on how to manage projects is covered in the chapter on Project Management (Chapter 4). When the components are about to close, the Program Manager needs to give approval for their closure. The Program Governance Board might also be involved in approving the component closure.

During the component closure, the PgMO, along with the Project Management Office (PjMO), can work together to organize and conduct component lessons learned meetings so that the knowledge transfer can occur. Usually, the Center of Excellence (COE) is involved in abstracting the lessons and creating a lessons learned report (or similar artifact), which is of interest to future similar projects.

Benefits realization is another sub-phase that is part of the program execution phase. Benefits are owned by the concerned Functional Managers in the areas in which they accrue. For instance, an automotive manufacturer may envisage increasing the market share of its company from 10% to 15% in three years (which represents a benefit toward achieving its strategic objective of being a market leader in the geographies in which they operate). Multiple projects can be launched by the manufacturer to increase this market share—such as launching a new car model, developing marketing partners, and maintenance arrangements, etc. The head of sales and marketing departments can claim the benefits of this program. As reiterated earlier, benefits management lies at the core of program management. Whereas portfolios can launch programs, it is the realization of benefits that defines the success factor for the program.

The Functional Managers alone cannot realize all the benefits. The functional teams can support benefits realization by working closely with concerned Component Managers. To illustrate:

- The Manufacturing Head of a company wishes to increase the volume of products it produces by deployment of Information Technology (IT) applications through optimal production scheduling systems.
- Multiple Project Managers work on the IT applications to develop a production scheduling system, which will increase the volume of products.
- The functional teams from the manufacturing department work along with these Project Managers to give user specifications for the production scheduling system.

- The Project Managers (and the project teams) develop the applications for which the User Acceptance Testing (UAT) is carried out by the functional teams. The final acceptance can be approved by the concerned Functional Manager (which, in this case, is the head of the manufacturing division).
- It is the responsibility of the concerned Functional Manager to prepare the operational units for the change (in this case, the manufacturing department users, who need to be willing to accept the change to handle the redesigned applications).

Whereas in this particular case, the change may not be of a sweeping nature, other transformational programs can induce far more changes, creating uncertainty and trauma for the functional departments. Usually, the Program Sponsor needs to give approval for the commencement of the transition, as any false moves here can create organizational chaos. The concerned Functional Manager may make recommendations to the Program Sponsor regarding this transition after assessing the organizational readiness for change and ensuring that the baseline values of the benefits are captured to measure improvement.

We cover organizational change management in Chapter 6.

Before transition, the Functional Manager will also create a "Benefit Card" to capture the relevant information concerning the benefit. This card can include particulars, such as the benefit identifier; benefit description; base value of the benefit and its expected trajectory across the benefit lifecycle; components and outcomes, which are needed to realize the benefits; benefit realization schedules and costs; associated risks; and the name of the benefit owner. The benefits realization plan is also created indicating the sequence of likely realization of the benefits and outcomes.

Once the UAT is completed, transition management to the operational units can commence.

When the actual system goes live, the impacted users get to know the real implications of using the system. There could be resistance to change that may need to be managed.

It may be necessary to have parallel runs until the new systems stabilize. Temporary facilities and teams for helpdesk management and user support would need to be in place during the transition.

A program can have multiple transitions during its lifecycle, and they can be testing times for all concerned stakeholders. During a particular transition, once the capabilities are delivered to the relevant functional department, the concerned Project Managers can be disengaged, and the Program Manager can focus on moving into the work for the next transition, after the corresponding outcomes have been stabilized. The access to legacy systems can be disabled once the transition has been completed.

During post-transition, benefits begin to be realized. The benefit card and the benefits realization plan are updated by the concerned Functional Manager. The progress of achievement of the benefits is monitored, and in case extreme deviations are noted, the Functional Manager needs to escalate them to the Program Governance Board and also to the PMG, as needed. It may be necessary to launch additional projects to stabilize the outcomes and benefits, which could be done as a part of the ongoing program. During transition management and benefits realization, the Functional Managers need to ensure that the quality of the service provided to the performing organization (and to external clients, if applicable) is not impacted beyond acceptable levels.

3.11 Program Closure Phase

After all components of the program are successfully completed and transitioned, the program itself can close, as a part of the program closure phase.

During normal program closure, the Program Manager needs to produce an End Program Report for the approval of the Program Governance Board. Normal closure can commence once the last set of components has been transitioned to the BAU. The end program report may contain the following information:

- Confirmation from the Program Manager that all components have been transitioned
- Confirmation from the concerned Functional Managers that the intended outcomes have been achieved and benefits have begun to accrue (or a specified extent of benefit has been achieved)
- Arrangements for transfer of ongoing contracts to the BAU
- Arrangements for transferring ownership of pending risks and issues (which are not expected to be major ones during the normal program closure)

- The Program Manager's own assessment on how the program went (including lessons learned report and feedback to Corporate strategy)
- A program wind-down plan, covering how the remaining resources will be returned, disbanding arrangements for the PgMO and archiving plans for the program assets

As stated earlier, it is not necessary that the full extent of benefits be realized as a part of the program lifecycle. As part of the program closure criteria, the Governance Board may stipulate that the program can be closed if it realizes, say, 80% of the value of the intended benefits, and the remaining benefits can be realized by the BAU.

In many cases, the Program Governance Board may seek the approval of the Portfolio Steering Group for program closure. This is more appropriate for the programs that were launched to achieve critical strategic objectives.

Premature closure of the program is also possible. This can occur for multiple reasons—a trigger from portfolio management, the business case becoming unviable, changes in key stakeholder support, withdrawal of funding, critical issues and risks affecting the program, etc. In all cases, the Program Manager prepares the end program report and seeks the approval of the Program Governance Board for early closure. In the case of premature closure, ongoing projects are handed over to the BAU (or grouped as a part of a new program, if need be, by the Portfolio Steering Group).

Concerned stakeholders are informed about the program closure, and a financial closure date for the program is normally set, as well. It also becomes the responsibility of the Program Manager to complete the performance appraisals of the core program team members and provide reverse feedback to the Human Resources (HR) Department or the Resource Management Group (RMG) on ways to upgrade skills, as needed. A complete review of the program is commissioned by the Program Governance Board, as well, to assess the extent of realized benefits to report to the PMG and the Portfolio Steering Group. Once the report for the program closure is accepted by the Program Governance Board, the program can close. Further work to sustain the benefits, reviewing the extent of ongoing benefits realization, and escalating to senior management if the benefits go off track, etc., are addressed by the concerned Functional Managers.

Chapter 4

Project Management—Delivery Enabler for Change

4.1 Project Management—Context

In the previous chapters, we discussed the concept of portfolio and how it spawns programs. We also studied how projects get started as program components. Projects are the fundamental enablers to effect change. It is noted that organizations having a higher project management maturity are able to drive change better to realize outcomes and achieve their strategic objectives.

As in programs, projects are also "temporary" endeavors. Typically, projects run for months, whereas programs run for years. However, the fundamental difference is that, whereas programs are undertaken to realize outcomes and benefits, projects are taken up to produce deliverables (which we call outputs in our discussion). A project can also be directly linked to the portfolio.

Projects have their own lifecycle, and each project can be divided into multiple stages. Each stage boundary can have a tollgate output (or a group of outputs), which needs to be approved by the Project Review Board (PRB) before going further on to the next stage. The personnel represented in the PRB can include the Project Sponsor (who could be the Program Manager for critical projects in the program) and representatives of functional departments who would be testing and accepting the outputs from the project.

Different frameworks, such as *A Guide to the Project Management Body of Knowledge* (PMBOK® Guide, 5th ed.),* use varying nomenclature for the processes in the lifecycle of the project. In this book, we will use a distinct nomenclature for describing the processes in the project lifecycle, as presented below.

4.2 Project Management—Major Processes

The major processes in a project lifecycle include initiating, setting up, delivery and monitoring, and closing. All these processes recur across multiple stages of the project.

If the project is a component of the program, the project charter can be issued by the Program Manager or Project Sponsor (who may report to the Program Manager). If the project is directly under the portfolio, senior management can issue the project charter.

4.3 The Project Charter

The project charter contains the following information:

- Name of the project (some projects give nicknames or codes, for easier identification)
- Name of the Project Manager and the Project Sponsor
- Project objectives (i.e., macro-level scope, schedule, and budget estimates, including any tolerances assigned to these and acceptance criteria for the project outputs)
- The Outline Business Case (OBC) for the project. (This can include information on how the project is linked to the strategic objectives of the organization, and how it will lead to the outcomes and benefits

* Published by the Projected Management Institute (PMI).

envisaged by the program, if the project is a part of the program. This information could be added by the Program Manager.)
- Description of major stakeholders and their expectations from project outputs
- Major risks for the project (and an overview of the likely risk responses)
- External dependencies of the current project with other components
- Other constraints/assumptions concerning the project
- Quality standards and guidelines to be followed by the project (especially relating to Corporate- or program-level guidance)
- Progress governance, reporting, and escalation mechanisms from the Project Manager to the Program Manager or to the PRB.

For the projects controlled by the program, the project charter is an output of the initiating process and can be refined during stage boundaries of the project's lifecycle. For the projects directly under the portfolio, a project mandate (or a feasibility study) can become an input to the initiating process. The mandate can provide high-level information on the expectations from the project, governance arrangements to be followed by the project, etc.

The very step of preparing the project charter enables consideration of the viability of the project and facilitates closer scrutiny of "pet projects."

The project charter needs to be approved by the PRB before the project moves onto the project setup process. The PRB looks specifically at the viability as assessed from the OBC, alignment of the project with program (or portfolio) objectives, assessment of major project risks, and achievability of the project objectives before giving the approval. As stated, if the project is part of the program, this approval can come from the Program Manager or the Program Governance Board.

The project charter empowers the Project Manager to requisition necessary resources for further planning. The level of planning at the project level gets into more detail as compared to the program-level planning.

4.4 Project Stakeholder Engagement

As in program management, the initial step the Project Manager needs to take is to identify the project-specific stakeholders, their stances, interests, etc. Whereas the Program Manager can address major stakeholders (and those stakeholders cutting across multiple projects), it becomes the specific responsibility of a Project Manager to identify the project-specific stakeholders and address their needs. At the project level, these stakeholders can include:

- Project Team Managers/team members
- Vendors specific to the project
- Customers (for external projects) and concerned functional departments (for internal projects)
- Project Review Board/funding agencies (which can also be addressed by the Program Manager)
- Public/regulatory agencies, etc. (if applicable)

A separate section on stakeholder engagement (see Chapter 6) discusses how to identify, segment, and address the stakeholders (including at the project level). The concerned project-level deliverables include the project stakeholder engagement plan and the stakeholder register. The project Communications Management Plan is intricately linked to the project Stakeholder Engagement Plan. The communications management plan includes details regarding the following:

- Who are the critical stakeholders and what types of communications need to go to them?
- Who are the senders and receivers of various communications?
- Which level of reports needs to go to the stakeholders and what is their periodicity and format?
- What communication channels are to be used (unidirectional or bidirectional and the channel descriptions)?
- How are stakeholder concerns to be addressed?

The structure of the communications management plan can align with Corporate standards as applicable. The level of scoping done at the project level is more detailed as compared to the program level.

Usually, the Project Manager sets out a scope management plan, which provides guidelines on how to gather requirements, develop a scope statement, create the detailed Project Work Breakdown Structure (PjWBS), manage scope changes, and obtain user acceptance. The scope management plan is integrated into the Project Management Plan and is developed as an output of the project setup process.

The stakeholder register developed during the initiating process is expanded when more stakeholders are identified during the requirements elicitation. For internal projects (projects that are done within the organization), the requirements gathering process is somewhat easier, as most of the stakeholders are delineated by their designations, and the Project Manager is usually aware of the relative importance of such stakeholders. The Project Sponsor can also facilitate the requirements management process, especially relating to senior management. For external projects (which are projects executed by a third-party company, usually under a contract), the requirements gathering can be more daunting, as the Project Manager may not be fully aware of the stakeholder interests and their requirements during the commencement of the project. Various techniques enumerated in the chapter on Stakeholder Engagement are useful here.

4.5 Requirements Management

Requirements management is a key to project success. This includes a structured approach for gathering requirements, determining their prioritization, managing changes to requirements, and ensuring that the requirements are implemented appropriately in the final deliverables. In industries requiring compliance, it is also necessary to demonstrate that the project deliverables meet the statutory and mandatory requirements. This aspect is more important for projects, for example, in finance, healthcare, and defense. With a proper requirements management implementation, development costs can be reduced by up to 57%, time to market can be accelerated by up to 20%, and the cost of quality can be lowered by up to 69%, helping to reduce the amount of rework with an accelerated time to market.* Poorly defined requirements are the major cause of rework, which gets perpetuated through uncontrolled requirement change requests, especially in software development projects.

The "right" requirements enable release of right products faster. Requirements management becomes even more complicated in situations in which diverse stakeholders have their own agendas and would like to see that they get priority during implementation. We covered some of these aspects in our discussion about portfolio management as a

* *Source:* IBM; http://www-03.ibm.com/software/products/en/category/SW740.

part of initiative prioritization. Requirements management thus gets into "political turf wars" between various key stakeholders, during which time the Project Manager needs to seek the guidance of the Project Sponsor. The issues that normally arise during requirements elicitation include the following:

- Key stakeholders are not able to articulate their needs appropriately, mixing what needs to be done with what outcomes are needed to be fulfilled.
- Duplicated and inconsistent requirements arise from varying layers of management.
- Poorly communicated end user requirements are changed during the project lifecycle.
- Incomplete requirements progressively becoming clearer during the project lifecycle, adding to rework and overrun of schedules and efforts/budgets.
- Middle management is not clear about the strategy of the organization and, thus, is not aligning project objectives with strategic goals.
- Poor specifications in engineering development projects become evident (one of the root causes of product failures and not being able to address business needs).
- More importantly, when requirements change during the course of the project, a full impact analysis is not done on the other impacted requirements, leading to inconsistencies and product failures.

In Information Technology (IT)–related projects, the value of an effective requirements management system is even more critical. Typically, it has been the experience working with IT projects, that end users get more "ideas" when they see a prototype of a working model. Such cascading requirements put a huge strain on the developers that, once it gets out of control, derails the implementation schedules. IT companies typically place more attention on "Requirements Engineering," focusing attention on changing requirements and linkage/traceability across the requirements management lifecycle. Agile and Scrum methodologies can address the need to take care of evolving requirements during the project.

There are well-published techniques available for requirements elicitation. We will consider three well-known techniques that progressively clarify the requirements and enable the acquisition of a collective buy-in across the stakeholder community.

1. Interviews are usually one-one-one. These provide an opportunity to get the users' perceptions on the current status and what their expectations are. Interviews work well, especially with Senior Managers who would like to voice their views privately and not bias others' judgments, as well, during a group meeting.
2. Once the requirements are gathered in "one-on-one" interviews, they are deliberated in "focus groups," which typically consist of members from the same department/unit (including the members who participated in the interviews earlier). These discussions enable consideration of those requirements at the "focus" level that require reconciliation and reprioritization.
3. The last technique that can be used here is the "facilitated workshop," which is typically a joint, cross-functional gathering that includes key participants from different functional departments. Such workshops discuss those requirements that require resolution through joint participation.

Although it might seem to be time consuming, the above three-step model goes a long way in getting the key stakeholders' inputs before going further into detailed planning, and reduces later rework.

4.6 Project Work Breakdown Structure (PjWBS)

The scope management plan can also contain the requirements management plan, which provides details on how requirements are elicited, consolidated, prioritized, etc.

Whereas the requirements management and building scope can be perceived as a "bottom-up" approach to the building of the scope statement and the PjWBS, another technique is to start from the Program Work Breakdown Structure (PgWBS) and "drill downward" to determine the project-level deliverables. This "top-down" approach is more amenable when the project is part of the program, and the Program Manager takes the lead in determining the scope of the project. Some companies call this a Product Breakdown Structure (PBS), containing the list of deliverables to be produced by the project. An example of a PBS is presented in Figure 4.1.

This figure shows a "miniature" PBS for an aircraft, showing some of its elements. The codes (such as A, A.1, etc.) allocated for various elements

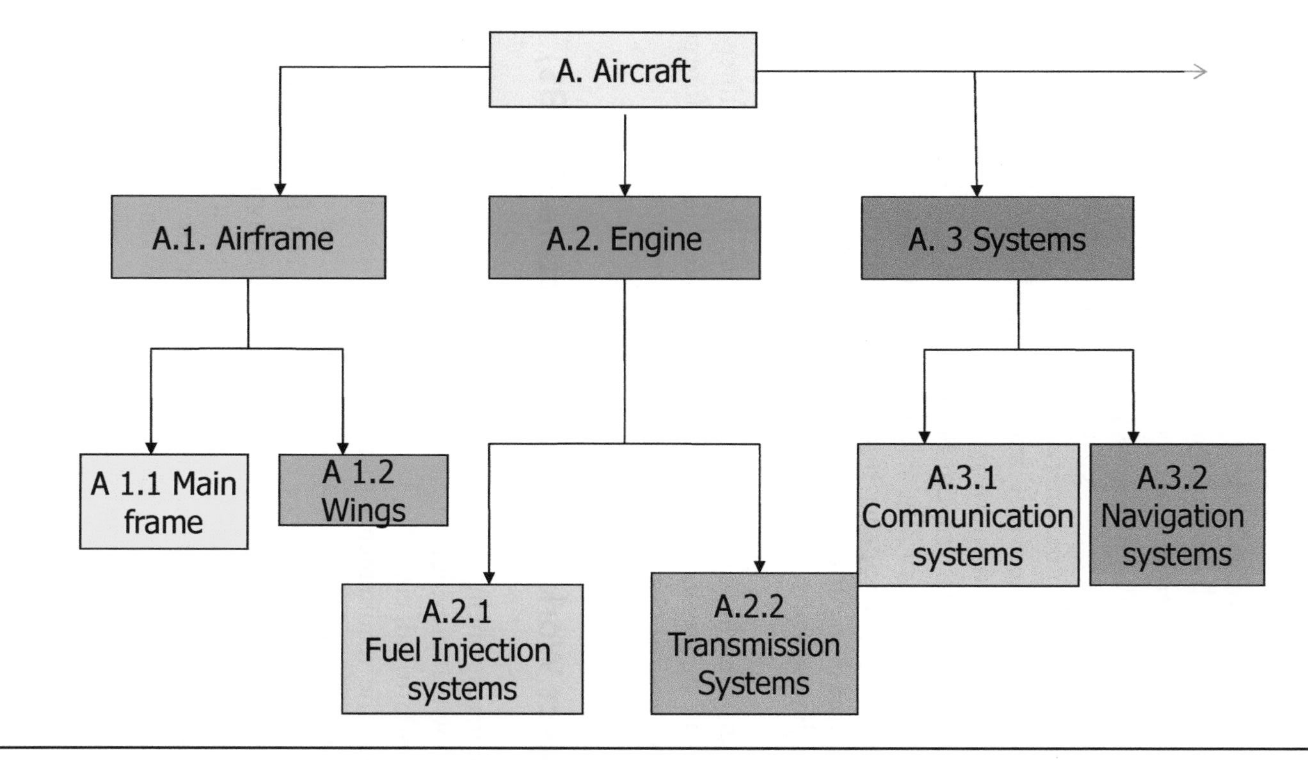

Figure 4.1 Product Breakdown Structure (PBS).

of the PBS are determined by the configuration management system and represent the unique product codes for the products concerned. In a program environment, the top level of the PBS can refer to the corresponding PgWBS Work Package. In this way, the linkage between program-level deliverables and project-level deliverables is strengthened.

As was stated in the chapter on Program Management (see Chapter 3), the advantage of creating the PBS (and the corresponding structure—PjWBS) is to provide clarity about what needs to be produced first rather than moving into questions about when they need to be or how they need to be produced.

It should also be noted that the entire PjWBS is not produced during the first stage of the project. It is quite likely that the first cut PjWBS contains the "macro-level" deliverables and the necessary drilldown is done only for those deliverables to be produced in the first stage of the project. The PjWBS represents a tree type of structure, consisting of both technical and management deliverables. During the project lifecycle, the work to be done for the next stage is considered, and the PjWBS elements for the succeeding stages are produced during the closing process of each stage. This progressive decomposition illustrates the "necessary and sufficiency" principle to consider only those deliverables that need to be produced during the current stage.

Typically the project-level WBS can be produced using two types of decomposition—sub-product-wise or stage-wise. In a sub-product-wise decomposition, the major product is decomposed into sub-products, which could be developed in parallel and could be in different geographic locations. Similar to an aircraft, the engine could be developed at one location, the wings could be developed in another location, etc., and they are produced in parallel and integrated together at a single place. In a stage-wise decomposition of the PjWBS, the products to be delivered during a particular stage are identified in greater detail. The details of the products developed during subsequent stages are identified just prior to the closing process of the preceding stage. Therefore, the project team is not overwhelmed with the need to develop product descriptions for all the deliverables and can focus on the current and the succeeding stage deliverables. However, how this technique is adapted to a project depends on the lifecycle model adopted (such as Waterfall/Agile, etc.)

The Project Manager develops a configuration management plan to identify and protect all the project-related assets. This plan indicates how the configuration items (the lowest-level deliverables) will be identified, stored, and archived. For each of the deliverables to be produced as a part of the project, a Configuration Item Record (CIR) is maintained by the Project Management Office showing the description of the product, current status (in terms of its completion), linkage with other artifacts, current product owner, and cross-references to concerned issues and risks. The collection of CIRs is an important reference document for the Project Manager that indicates which deliverables are under which stage of development and serves as a pointer to update the project management plan. In effect, the CIR is the most granular artifact in a project environment, describing all that needs to be known about a product, its delivery status, and interlinkages with other project-related products.

4.7 Project Schedule Development

After the deliverables have been identified, the next logical step is to determine when these deliverables need to be produced. This depends on multiple variables, including the effort required to produce the deliverables, the capability and availability of resources required, available budgets, and externally determined milestones, etc. As was noted in the chapter on Program Management (Chapter 3), up to the delineation of the deliverables, the project management lifecycle sequence is fairly linear. However, from then onward, determination of the schedule and costs is done iteratively until acceptable baselines emerge. In the following, we consider salient aspects regarding the determination of schedules and costs for the project.

Determination of the project schedule commences with an assessment of the efforts required to produce the deliverables. It should be noted that there are two types of deliverables in a project—technical deliverables (which are mostly produced by the Team Managers or the Vendors) and the management deliverables (such as the scope management plan, etc.), which are predominantly produced by the Project Manager. The PjWBS needs to include both of these elements since it represents all of the work of the project.

For each of the deliverables, the tasks necessary to produce these deliverables are identified. Efforts (could be in person-days) are estimated for the tasks and correlated with the availability of resources to calculate the calendar durations in which the tasks can be performed.

The sequence in which the tasks are performed is determined in parallel. Combined with the calendar durations in which the tasks are performed, the first cut project schedule is determined.

External dependencies (which are dependencies on external events not under the control of the Project Manager) can add more risks to the project.

The project critical path is determined next. The critical path is the "longest path" from beginning to end for a project and is the minimum time that will be required to complete the project. We depict the computation of the critical path by means of a numerical illustration.

In this diagram (see Figure 4.2), there are six tasks—A to F—with the associated durations (in weeks) provided. Table 4.1 illustrates the computation of the critical path and float time available for this diagram. In this table, the first two columns represent the name of the task and its duration. The third column represents the Early Start (ES) of a task, indicating how early it can start. Task A can start immediately (at the end of week 0). Thus, the ES of task A is 0.

The early finish (EF) of a task is ES + task duration, so EF for activity A works out to 2 (indicating task A can finish at the earliest at the end of week 2). Using the same logic, the ES for task D is 0, and the EF is 6. Now task B has two predecessors (A and D), and both of them need to be completed before task B can commence. Task A has EF as 2, and task D has EF as 6. Since both of these predecessors need to be completed before B can commence, the ES for task B is 6, and its EF is 10. Using the same logic, the ES and EF for all the tasks are computed from "left to right," as in a forward pass. The EF for the last task in the network diagram F is 14, indicating that given the topology of this network, no Project Manager would be able to complete the project in less than 14 weeks, with the given set of resources.

In the backward pass, the navigation occurs from right to left, commencing from the last task. Thus, in the current case, for task F, the EF is at the end of 14 weeks. Assuming the deadline for completing the project is also 14 weeks, the Late Finish (LF) for task F comes at the end of 14 weeks. The Late Start (LS) for task F becomes 13 weeks.

Task C precedes task F, and, hence, the LS for F becomes the LF for C (at the end of the 13th week), and the LS for C becomes 13–3 = 10th week.

For task E, there are two successor tasks, C and F. The LS for task C is 10, and LS for task F is 13. The LF of task E ought to be the minimum of

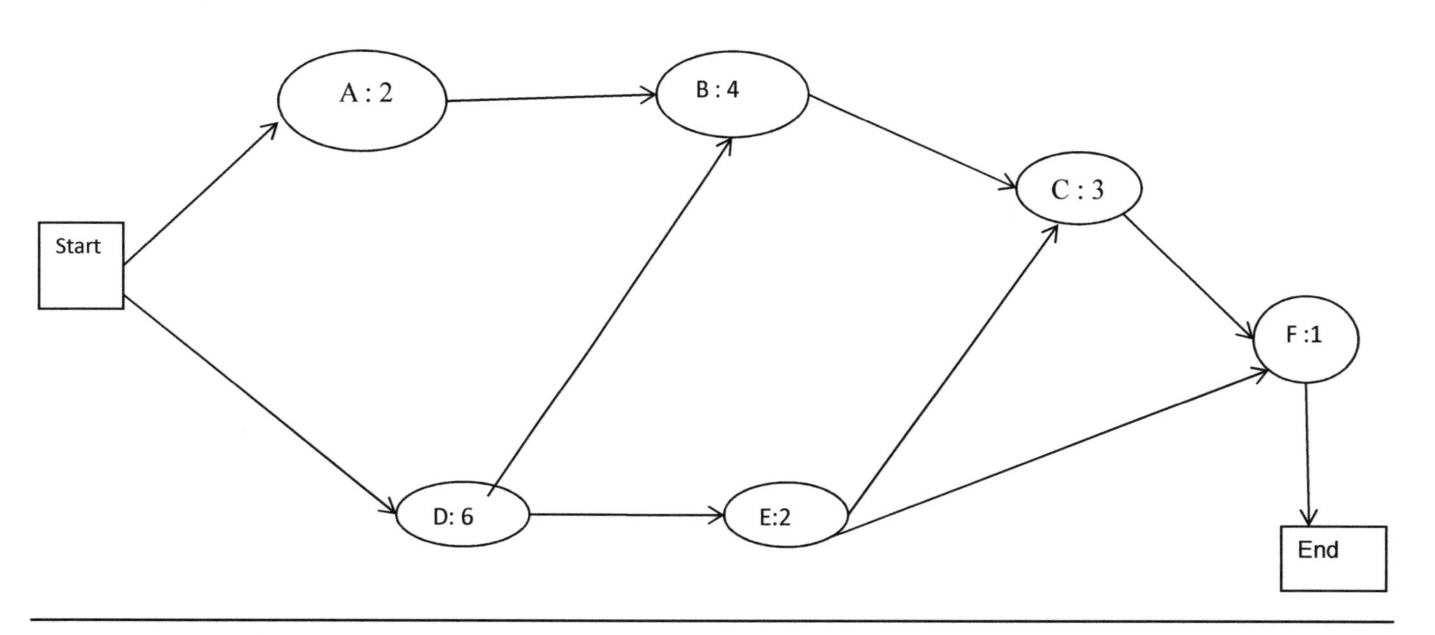

Figure 4.2 Critical path diagram.

Table 4.1 – Critical Path Computation—An Illustration

Task	Duration	Early Start (ES)	Early Finish (EF)	Late Start (LS)	Late Finish (LF)	Total Float (LS-ES)
	(in weeks)	(at the end of the week)				(in weeks)
A	2	0	2	4	6	4
B	4	6	10	6	10	0
C	3	10	13	10	13	0
D	6	0	6	0	6	0
E	2	6	8	8	10	2
F	1	13	14	13	14	0

these (which is 10), and its LS then becomes 8. Using the same logic, the LS and LF of various tasks are computed in the backward pass.

The last column in Table 4.1 represents the "Total Float," which is computed as LF-LS (which is algebraically the same as EF-ES) for all the tasks. The critical path is a logical sequence of tasks having a zero total float. In the present diagram, it works out to the path D-B-C-F, and the critical path duration is 14 weeks.

The total float gives further information on what can happen if some tasks are delayed (say, due to non-availability of resources). If the tasks in the critical path are delayed, the total project will also be delayed. However, there is some leeway for the noncritical tasks. For instance, for task E, the ES is at the end of the 6th week, but it can be delayed up to the end of the 8th week without affecting the total project duration of 14 weeks. These insights enables the Project Manager to allocate resources to the critical path tasks based on priority, even by withdrawing from noncritical tasks (if feasible). It should be noted that when the actual resources (by name) are allocated by the Resource Management Group (or HR or functional departments, depending on the type of the project), the critical path could change, depending on the skill set or the extent of resources allocated. Here, we are predominantly considering the human resources needed for the project, but it can include other resources (such as the availability of hardware/network bandwidth, etc.) that can also constrain the project.

4.8 Project Cost Baseline

The cost of a task is predominantly driven by the rate of the resources needed to complete the task. Some of the resources could be commodities (e.g., cement/steel in a construction project) or services (e.g., the services of a janitor to complete a task). These resources could be internal or external. For internal resources, the organization may decide to impute an internal transfer cost to charge the project. A guidance document known as the financial management plan is usually prepared to assist the Project Manager in determining how to allocate costs for such resources. The financial management plan (or Cost Management Plan, as it is called by some standards) can include the following information:

- How to allocate costs for various types of resources
- How to roll up the costs from individual tasks up to the project level
- How to account for the actual costs of the project (especially in certain type of industries, inventory management techniques such as First-in-First-Out or Last-In-First-Out need to be applied for goods supplied for the project work, although it is more applicable for operations)
- How to account for cross-currency exchange rates in case the project is being executed in different countries
- Project cost reporting in case of financial year closings of various companies, etc.

Costs of the projects are multifarious. These can include:

- Costs of the development of technical products (which are predominantly incurred by the vendors and the Team Managers or members)
- Cost of developing management products (such as the risk management plan, project progress reports, etc.), incurred mostly by the Project Manager
- Contingencies, as needed

How to develop contingencies is more a risk management concept—to address "known-unknowns." Usually, companies allocate 10%–15% as a contingency for various tasks, but it depends on the type of project. Research and Development (R&D) projects could get more reserves, whereas routine maintenance types of projects may get a lower contingency reserve. Past information on project performance furnished by the

Project Management Office can be useful here. The overall cost baseline is often referred to as an "S-Curve" and becomes a reference point for measuring deviations during project execution.

4.9 Scope Change Requests and Managing Change

Once the project management plan gets baselined at the end of the project setup process, issues can arise during the project delivery and monitoring process. Typically, there are two types of issues identified—change requests (which call for a change in scope from what was baselined) or deviations (the inability to fulfill the agreed-upon scope, etc., because of intervening factors). The requests for change usually come from the user community, and the deviation requests (which may also be called off-specifications) come from the supplier community. In either case, such issues need to be analyzed by the Project Manager during project delivery, and corrective and preventive actions need to be taken, as appropriate. For some projects, a budget for an upper limit for change requests is set, which is jointly agreed upon by the end users and the Project Manager. This is a good practice, as it inherently places a cap on the extent of change requests that can be raised by the user community and enables them to think prudently before suggesting change requests.

It is quite likely that the project has billable milestones (especially for external projects), and the spending in between billable milestones can exceed the intermediate funding. It becomes the responsibility of the Project Manager to reconcile these mismatches between expected spending and expected intermediate fundings by, for example, rescheduling procurements. Such reconciliations may have an impact on the project's critical path.

As stated earlier, in a project environment, scope, schedule, and cost form an "iron triangle." If one side of the triangle is changed, it will have an effect on the other sides of the triangle. Hence, all three variables need to be managed synchronously by the Project Manager.

4.10 Project Quality Management Plan

As compared to the program level, quality in the project level is more product focused. Therefore, the Quality Management Plan produced at the project level will include more details on the following:

- Quality management standards to be followed for the project (which may include statutory guidelines and compliance standards to be followed)
- Roles and responsibilities for ensuring quality of deliverables
- Quality management procedure to be adopted for the project deliverables (such as peer reviews, testing procedures to be adopted, etc.)
- Tools and techniques that could be adopted for quality management. These can include statistical tools, including control charts and diagramming tools such as fish-bone diagrams, Pareto charts, etc.
- The quality records that will be maintained for the project, who will undertake quality audits and reviews, etc.

The project quality records need to be integrated with the configuration item records, and the project quality management plan needs to interface with the project governance plan (which states how the governance processes in the project will be invoked).

For external projects, blending the quality standards of the client and those of the supplier may be required. The lifecycle adopted for the project may decide which types of quality reviews may be adopted. For instance, Agile projects may follow a different lifecycle as compared to traditional Waterfall model projects. In Agile projects, timeboxes are usually created, and the deliverables are prioritized on a "Must be done, Should be done, Could be done and Will not be done" basis (usually abbreviated to MosCow analysis). Within each timebox, the "Must be done" and "Should be done" deliverables are given priority for delivery. The quality management plan can also include references to the roles of external parties, such as auditors, reviews by funding and statutory agencies, etc.

4.11 Project Communications and Risk Management Plans/Risk Management Flow

Communication and stakeholder engagement plans are quite similar to what we discussed briefly as a part of program management (Chapter 3), and they will be reviewed again in an exclusive chapter (Chapter 6: Stakeholder Engagement).

The risk management plan at the project level focuses on what needs to be done to address risks specific to the project. If the project is a part of the program, much of the guidance will be given by the Program Manager to ensure uniformity across the projects. The sources for the project-level risks mainly come from the team-related deliverables, skill sets, resource availability, vendor management, scope creep, etc. The risks can also emanate from the programs (as they could relate to funding, inter-project dependencies, other external dependencies, etc). The risk management plan at the project level can provide guidance on how to identify the risks, prioritize and address the risks, escalate the risks as needed, etc. The risk management cycle in a project goes as follows:

- The risk management plan provides guidance on the risk management cycle to the Project Manager.
- The Project Manager opens up the Project Risk Register and identifies relevant risks, in consultation with SMEs, top management, and even the Client, if applicable. Multiple techniques can be deployed for risk identification, including group discussions, analysis of past lessons learned, Root Cause Analysis (RCA), etc.
- Many companies create a Risk Breakdown Structure (RBS), correlating the possible risks with the sources of the risks. The Program Management Office can facilitate this taxonomy classification.
- The advantage of this grouping is it enables the risk sources to be considered when addressing the risks for effective treatment.
- Anyone can identify a risk based on their past experience, knowledge, etc. The person identifying the risk becomes the risk author.
- Each risk is allocated to a "risk owner"—the person or the role best suited to address the risk. The description of the risk owner can vary with the type of risk. For vendor selection–related risks, the procurement department can become the risk owner. For team skill-related risks, the HR department or the Resource Management Group (RMG) can become the risk owner. These are the roles that are in the best possible position to address the risks because of their in-depth knowledge about them.
- Once the risks are identified, they are entered into the project risk register. The Project Management Office can facilitate the opening of the blank risk register and assist the Project Manager in the administrative tasks concerning the risk management cycle.
- The risks need to be noted systematically, including the risk cause, the risk event and the likely risk response (which can be completed later). The risk cause is usually something that is already known (e.g., constructing a building in an earthquake-prone area).

- The risks are next prioritized on three parameters—probability, impact, and proximity (stating when the risks are likely to occur first in a project lifecycle)
- Many companies adopt a two-step prioritization—qualitative (or subjective) followed by quantitative. In qualitative prioritization, normative scales are used (for instance, for probability they could be low, medium, or high or on a scale of 1 to 10) and a similar classification is used for the impact. The advantage of performing a qualitative risk analysis initially is that it enables an opportunity to take up a first cut prioritization of the risks and focus on high-priority/high-impact risks for a further detailed analysis as a part of the quantitative risk management. The probability-impact (PI) matrix used for the program management risk classification is also (and, in fact, to a greater degree) applicable at the project level. The risk owners, along with the Project Manager and other SMEs, should be part of the prioritization of the risks. It needs to be noted that this prioritization is done on an ongoing basis during succeeding delivery stages.

Once the risks are prioritized, the next logical step is to address the risks. Broadly, the following steps can be taken during risk response planning.

- Avoid the risk if it is feasible: This approach calls for eliminating the root cause of the risk so that the risk cannot occur, or it becomes "irrelevant."
- Transfer the risk to a third party that can better address the risk. The third party bears the impact of the risk, for which money needs to be paid to them. Insurance, warranty, outsourcing the application support with predefined service-level agreements (SLAs), etc., fall under this category. By transferring the risk, the probability of the risk occurring is not reduced; only the impact is.
- Mitigate the risks, by taking proactive steps to reduce the probability and the impact. Testing a product before a customer release is a classic example. Mitigation also incurs costs that need to be weighed against the losses, which can occur due to the release of faulty products, etc.
- The final risk response measure is accept. There are two sub-responses here, including active acceptance (which includes preparing a contingency plan, but activating it when the risk is likely to occur) and passive acceptance (which is essentially a "do-nothing"

option). Usually, passive acceptance is invoked for minor risks or risks against which no other immediate response is possible.

Risks could be positive as well (relating to opportunities). The responses to opportunities could include the following:

- Exploit—make the probability of the opportunity occurring closer to one through better management. This approach can include, for example, putting in the best prototypes, etc., to get a client contract.
- Share the opportunity with a third party who is better equipped to address it.

To illustrate, while expanding the business in a new territory, set up an alliance with local providers in the geographical area to enable better insight into the market.

- Enhance the opportunity by increasing the probability and impact of the opportunity. Some marketing companies use this approach through cross-selling and up-selling multiple lines of products.
- Accept—do nothing.

The risk register is continuously updated during the project lifecycle. Typically, the project cannot get closed if a major risk is pending. It should also be noted that there is a tight link between the risk management lifecycle and the issue management lifecycle. If a company is not managing risks adequately, it is usually ends up with more issues. And issues can give rise to new risks.

The risk management plan can also specify when to perform risk audits to see if the risk procedures in the plan are being followed. This is important in high-maturity organizations to sustain competency in risk management.

The aggregate risk exposure for the project can be calculated as the sum of probability times the impact across all risks. The risk management plan can set the risk exposure tolerances for the overall project. It ought to be noted that the aggregate risk exposure keeps on changing over the course of the project and needs to re-evaluated at the end of each stage of the project. If the aggregate risk exposure crosses the risk tolerance, the project may become unviable and may be considered for closure. The aggregate risk exposure at the program level is derived based on the risk exposure of constituent projects and risks emanating from non-project work. Changing risk exposure for the project can influence the aggregate risk exposure at the program level.

4.12 Procurement Management and Staffing Management Plans

Large projects also have contracting needs. The Project Manager consults the procurement division in selecting the appropriate vendor. How to select a proper vendor comes under the realm of vendor management and is not discussed in detail here. However, it becomes the responsibility of the Project Manager to:

- Ensure that the appropriate vendor gets selected (more so for service-level procurements)
- Inform the development and documentation standards applicable for the project to the vendors
- Clearly define the scope of work to be performed by the vendors and specify their acceptance criteria
- Monitor the work of the vendors and their interactions with the rest of the team, as applicable
- Test and accept the deliverables produced by the vendor
- Ensure the payments to the vendor are remitted after the acceptance of their deliverables

Resources are assigned for the internally produced deliverables. Especially for the Human Resources (HR) allocation, a Staffing Management Plan is produced by the Project Manager. This plan contains the skill set of required resources and their likely deployment dates, the functional departments from which they need to be sourced, etc. The staffing management plan may also contain the reporting relationships within the project, which become part of the project management plan.

4.13 Project Setup End-Deliverable: Project Management Plan Finalization

The project management plan is developed at the end of the project setup process during the first stage and is updated during subsequent stages.

The project management plan provides a complete description of "what needs to be done in the project, how it has to be done, and when and how the project will be closed," and it becomes the master reference document for the Project Manager.

The project management plan will also include a project governance plan, which informs how the project progress will be monitored, how the Team Managers will be reporting progress to the Project Manager, issue- and risk-escalation procedures, a description of tollgate and other review procedures, and the project-closure procedures. If the project is part of the program, many of these guidelines will need to be aligned with the program-governance requirements.

At the end of the setup process for the first stage, the project management plan and a plan describing how to deliver the first stage outputs are placed before the Project Review Board for approval. The Board considers the following factors while giving approval to go ahead with the project (and this applies to subsequent stages during the project as well).

- Is the project aligned with the strategic objectives of the organization and the program goals (if the project is part of the program)?
- Is the project viable (from the business case perspective)?
- Are the targets achievable, given the capability and capacity availabilities?

4.14 Assessing Project Viability

Multiple techniques to assess the viability of the project are used by the Project Review Board. These techniques can include consideration of the Payback Period, Net Present Value (NPV), Internal Rate of Return (IRR), etc. All these techniques basically assess the likely costs and benefits of the project (imputed in monetary terms) and compare the costs vis-à-vis the benefits. With the NPV technique, the cost of capital (or a similar rate) is used to compare the discounted flows of costs and benefits to ascertain if the investment is likely to be profitable. With the IRR, no specific external interest rate is considered, but it is derived internally based on expected inflows and outflows.

For an illustration, consider Table 4.2. In this table, the expected incomes and expenditures are noted, as per discussions with business analysts and analysis of market projections. It is also assumed that the

Table 4.2 – NPV—An Illustration

Year	Expected income (in thousands of dollars)	Expected expenditure (in thousands of dollars)
0	—	100
1	50	20
2	70	10
3	50	10

expected interest rate is 10%, representing the cost of capital (which is typically the average weighted rate of interest to be paid to multiple funding sources, etc.). All income and expenses are expected to accrue at the commencement of the year.

During the year 0 (which is during the start), no income is expected, and the expected cash outflows are $100. Therefore, the net profit for the first year is expected at (–)$100. During the commencement of the first year, an income of $50 is expected, and an expense of $30 is forecast, yielding the net income projection of $20. Since this $20 is expected to accrue during the commencement of the first year, it needs to be discounted to the present rate. The discount factor is taken as 10%.

The discounting rate formula is PV = FV/(1+r)**n, where PV is the present value, FV is the future value of an investment, r is the discount rate and n is the number of years considered. (This is a variant of a compound interest rate formula.)

Applying the above formula, the NPV for the first year is 20/(1+0.1), which will work out to about $18. Likewise, the NPVs for each of the years is calculated and aggregated across the project lifecycle, which will work out to about $5.8. This implies that the project is financially viable with the given set of assumptions and data projections.

In IRR, the rate of interest (r) is taken as an algebraic variable and internally calculated to equate the discounted incomes to discounted expenses. Then, the "break-even r" becomes the IRR, which needs to be compared against the external "hurdle rate" to determine if the project is likely to be viable or otherwise. More advanced techniques are available for project accounting, which can be obtained from open sources. The project viability is assessed at the end of the project setup process and during the end of each of the stages to determine if the project should continue.

4.15 Project Delivery Process

Once the project management plan is approved by the Project Review Board, the Project Manager can focus on the delivery process. The delivery process consists of the following activities:

- Investigating appropriate lifecycle or development models applicable for each of the work packages to be assigned to Team Managers (who could be internal resources or external vendors)
- Discussing and assigning work packages to the Team Managers (including the description of the scope of the work package, envisaged schedule, budgets, quality and reporting standards to be kept in consideration by the Team Managers, permissible tolerances for the work packages, etc.)
- Obtaining regular progress reports from the Team Managers and updating the project management plan (the Project Management Office can facilitate this task if one is available)
- Dealing with issues and risks raised by the Team Managers
- Making sure that the team-level deliverables are up to the requisite levels of quality and are accepted by the concerned users
- Once the work package gets developed and accepted, allocating further work to the Team Managers (or reassigning resources, as needed)
- Motivating and encouraging the team to be at their peak productivity
- Giving feedback to HR or the Resource Management Group (RMG) on the skill sets acquired by the Team Managers and the gaps noted. Likewise, giving feedback to the procurement department on the caliber of vendors selected also occurs during this process.

In a typical project, about 90% of the effort and budgets are spent for technical deliverables that are produced by the Team Managers, who interact with the team members to get the work done. The Team Manager may produce his or her own team plans to monitor the work allocated to them. It should be noted that all the team plans may not follow the same structure, as some of the vendors can choose to produce the team plans as per their company standards, quality requirements for the products delivered, and the lifecycle models adopted. Nevertheless, it would be preferable for the Project Manager to align the progress reporting periodicities from various teams for easier updates with the project management plan (which also contains details of the project schedule and costs). The Team Managers also produce team progress reports to be given to the Project Managers. These team progress reports can contain the following information:

- Date of the report
- Follow-up from the work done during the last reporting period
- Work done during the current reporting period, including deliverables completed, in progress, and to be done
- Work forecast to be done in the coming reporting period
- Progress of the team plan (against allocated schedule, budget, etc.)
- Status of closed and pending issues and risks

4.16 Project Progress Monitoring Process

This project progress monitoring process is the interface between the Project Manager and the Project Review Board (PRB). This process is an overarching process, which "envelops" across the stage, overseeing how the stage is initiated, set up, executed, and closed. The progress reports sent by the Project Manager are reviewed by the PRB, and they can provide guidance to the Project Manager as needed.

This process also deals with escalations of issues and risks by the Project Manager to the PRB. The procedure for risk management was noted earlier. In the case of issues, the Project Manager performs the following steps:

- Records the issue in the Issue Register (or have it recorded by the Project Management Office).
- Assigns the issue author and the issue owner (definitions are similar to risk management).
- Prioritizes the issue (usually across the scales of priority of the issue and its severity). The Project Manager needs to enlist the assistance of the issue author/owner/concerned Team Manager and other Subject Matter Experts (SMEs) during this prioritization.
- Devises the issue resolution plan (which could include, for instance, increasing the budget, reducing the scope, removing unessential

features, etc.) along with the corresponding impact on the project management plan.
- The Project Manager could devise multiple alternatives for consideration of the PRB, which decides on the most appropriate course of action depending on the viability of the project, its criticality to achieve the business objectives, technical considerations, resource availabilities, etc.
- Once the PRB gives its decision, it usually becomes the responsibility of the Project Manager to accept the decision, change the project management plan and communicate the changes to all concerned (including the Team Managers, vendors, and impacted stakeholders, etc.), and re-execute the project accordingly.

The issue register can be maintained by the Project Manager or the Project Management Office. This register can maintain the record of each of the issues; their status and if closed; the closure date; and if the issue was closed, the issue closure date.

The Project Manager (or the Project Management Office) can open a log for capturing all the lessons learned. Anyone can provide inputs for lessons, which are scrutinized, classified, and recorded. If the organization has a Knowledge Management Office (KMO), it can facilitate the knowledge management process, with dissemination of the appropriate lessons learned to concerned stakeholders. The KMO can also facilitate the recording of best practices and metrics, which are useful when the organization is advanced in its maturity for handling projects and programs. The KMO can also be part of the Center of Excellence (COE), discussed earlier in this book.

The deliverables at a particular stage would need to be accepted by the concerned users. The information on who will do the testing, the testing procedures to be adopted, the deliverables acceptance criteria, etc., are usually prescribed when the work packages are allocated to the Team Managers. Depending on the deliverable, there could be multiple levels of tests, such as like unit tests, integration tests, etc., before the acceptance sign-off is given by the concerned users.

Once all the deliverables at this stage are completed, the stage-closing process can be invoked. The Project Manager can produce an "End Stage Report," signifying which deliverables were slated for development, the record of their approvals, any lessons learned, the extent of fulfillment of the business case, an updated project management plan, what comprised the deviation of the stage plan from what was planned, and any inputs concerning resource allocation and management. Along with this report, the Project Manager can produce the draft for the next stage plan, both of which go for approval by the PRB. The PRB approves these deliverables using similar criteria as was used during its approval of the initial project management plan.

After receiving the approval of the PRB, the project passes onto the next stage—the initiating process. Since the PRB has already approved this next stage, no surprises are expected when the Project Manager updates the project charter and sends it for further approval. However, the stakeholders for each stage could be different (or their influences can vary) so the Project Manager needs to update the stakeholder register and the activities for their engagement.

4.17 Project Closing Process

Once the end of the last stage of the project is reached, the "normal" closure activities can be invoked by the Project Manager. An "End Project Report" can be created by the Project Manager for the approval of the PRB. This report may contain the following fields:

- Reference code of the project
- Original and updated records of project management baselines (with a revision history thereof)
- An assessment of how the project went, including realization of the benefits, fulfillment of the business case, and achievement of its strategic objectives
- An assessment of what went right, what could have been done differently, and an extract of lessons learned
- Records of acceptance of all project deliverables (which are progressively obtained during the project lifecycle but reproduced more as a summary record)
- Acceptance of the overall project deliverable by the concerned user
- Handover and benefit sustainment plan (if applicable from the overall project deliverable)
- A review of team performance (including records of their performance assessment, if need be)

- Resource disposition arrangements (how the project resources will be returned to wherever they came from, including the closure of applicable contracts concerning the project)
- Deliverable archival plan—stating how the deliverables are archived for further scrutiny by the reviews, etc., as needed

The end project report also needs to be produced for "abnormal" closure of the project, which may be due to various triggers. These triggers can include withdrawal of funding for the project, the project business case becoming unviable, technical or resource-related issues, change in strategy from the portfolio or change in direction from the program, customer initiated closures, etc. During such an abnormal closure, the end project report will contain the status up to the stage where the project was abnormally closed. Any open issues and risks are usually handed over to the next project or to the concerned Operations Manager.

Once the end project report is formally approved by the PRB, the project can be closed. It needs to be noted for the projects in the program environment that it becomes the responsibility of the Program Manager to be involved in the project closure and to update the program management plan appropriately.

Chapter 5

Change Initiative Integration into Operations—Transition Management

5.1 Introduction to Transition Management

In the preceding chapters, we discussed how to design the portfolio and deliver programs and projects. In this chapter, we focus on integrating the changes into the operations of an enterprise. We also use the nomenclature Business As Usual (BAU) to connote the ongoing operations.

Many of the change initiatives are planned well, but execution remains a challenge. The integration into operations and realizing the outcomes and benefits is where the "rubber meets the road."

We will discuss the softer aspects of Change Management in Chapter 6. Here, we focus on the work that needs to be done (mostly by the Program Manager and the concerned Functional Managers) to ensure that the output of the projects and other work gets transitioned into BAU and that the BAU services are redesigned accordingly.

As we have noted in the chapter on Portfolio Management, every organization needs to balance its efforts, energy, and resources between planning and executing change initiatives and running the operations, which keeps the "lights on." It is the execution of operations that gives context and visibility to an enterprise. However, the organizations would also need to respond to external and internal triggers for change, which is facilitated by the change initiative management. Balancing resources between running the business and changing the business is an eternal "chase" for any organization.

Transition management comes into picture once the concerned outputs from the project have been tested and approved to "go live." Usually, the operations team that would take ownership of the outcomes would have been represented during requirements elicitation—guiding the Project Managers during product development, ensuring quality of products and services delivered, performing the user acceptance testing, and preparing the operational departments and customer groups for the change.

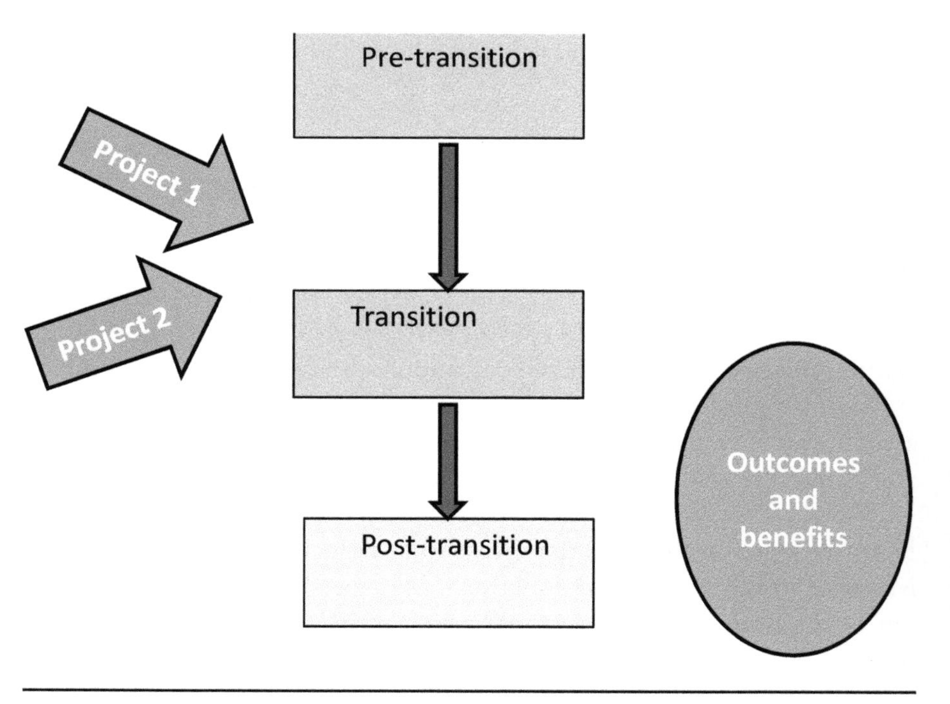

Figure 5.1 Benefits realization.

Broadly, the overall transition management can be divided into three steps, coming under the program execution phase and benefits realization sub-phase. These steps include:

- Pre-transition: Where the preparatory work is done by the business team regarding selling and leading the change to the business areas impacted
- Transition: Where the actual transition (and cutover) to operations occurs and outcomes begin to be realized
- Post-transition: Where the outcomes are stabilized and the benefits begin to accrue

A summary highlighting these steps is presented in Figure 5.1. Each of these steps is discussed in detail in the following sections.

5.2 Pre-transition Step

This step typically coincides with the project management lifecycle of the projects for each of the iterations of the program execution phase. (Please see the discussion in Chapter 3 on program management regarding the details of the program management lifecycle.)

The following work will be done by the functional/business area managers during this step:

- Preparing the impacted operational areas for change (further details are covered in Chapter 6 on change management).
- Updating the benefit card. The benefit card is defined by the structure stated in the section on Program Management (Chapter 3). It is quite likely that each iteration of the program executing phase maps to multiple benefits—in which multiple benefit cards can be prepared or updated.

The key fields to be updated in the benefit card include the following:

- The description of the benefit.
- The names of the benefit owners and a description of how the benefits and outcomes will be measured.

- The base value ("as-is" value) of the benefit. This is important because unless the base values of the benefits are captured, improvements cannot be measured accurately. To the extent feasible, the benefit measures need to get linked with existing outcomes so that they remain relevant to operations.
- The expected trajectory of the benefits across the benefit lifecycle. This could be described as, "the expected increase in market share of the product within first year of its launch is 3%, which is expected to go up to 5% in the second year and stabilize to 6% during the third year of its launch (after which a variant of the product may need to be launched to increase the benefits even further)."
- When the corresponding outcomes and benefit are expected to be realized.
- Linked project outputs and when these are expected to be handed over to operations.
- Any issues and risks concerning benefit realization.
- The change management plan for the impacted operational areas.

As was noted earlier, during the pre-transition step, the corresponding project outputs would be under development. It is imperative for the concerned Functional Managers (or their representatives) to monitor the project progress and prepare (or defer the preparation, in case of project time overruns) the impacted business areas for change.

The transition plan is prepared in detail during the pre-transition step. This plan includes the following information:

- When the transition will actually happen
- Prerequisites that need to be completed before the transition
- The skill sets of the people who will be assisting the Program Manager/impacted operational areas during transition, and how to deploy them
- A description of the temporary facilities required for the people who will manage the transition
- Arrangements for maintaining business stability during transitions (especially the service-level agreements (SLAs) with the clients of the impacted business areas)
- How the progress of the transition will be monitored

- Arrangements for parallel runs (to minimize discontinuity to business operations during transition) and also for rollback in case of any failures (especially for migrations in IT projects)

It is the responsibility of the Functional Managers to assess if the operational departments are ready for change. Typically, these changes are synchronized with BAU such that:

- They do not come immediately after a recent change that has not been fully absorbed by the business, in order to minimize "change weariness" of functional areas.
- These changes do not coincide with the peak operating seasons of the BAU. (As an illustration, for many retail selling companies, Christmas is a peak sales season, and they would not like to have a situation of unstabilized change just prior to this season. In the case of manufacturing companies, some of these changes in processes and systems are done to coincide with annual plant shutdown periods.)
- There is adequate support available from the BAU side during the transition – as it will call for additional resources to "double up" as Transition Managers in addition to their "day jobs."

Before the final "go-live" clearance, typically, the following criteria need to be met:

- The business users have tested the products/services, and user acceptance testing is completed.
- All major issues and risks concerning the products and transition itself have been addressed.
- The impacted business unit has created new roles, as needed, and filled them with their job profiles; finalized plans for office moves, if needs be; provided training for new procedures; and created necessary support arrangements.
- The temporary facilities for transition (including help desks/call centers, etc.) are in position.
- Necessary communication infrastructures, logins, and security procedures are created, as required, and the plans for transition are communicated to concerned stakeholders.

- In the case of Information Technology (IT)–related migrations, the necessary hardware/software/network infrastructure is installed and legacy data is "cleaned" and updated for completeness.

5.3 Transition Step

After the project outputs are tested and accepted, they are transitioned into live operations. Again, it is the responsibility of the Functional Managers to decide on the timing of these changes, as it is their business operations that ultimately will bear the brunt of these changes. The benefit card is also updated to finalize the timing of the changes, during which the following occur:

- During the actual "cutover," the functional resources will require support systems such as a help desk, "end-user" manuals, call center support, etc. These should have been planned beforehand as a part of the transition plan and are now put into effect.
- All major issues (and risks) from the outputs are addressed before the Functional Managers agree to take over the outputs. Reworks and retractions are always expensive and can undermine the credibility of change. The Program Governance Board, therefore, needs to approve the timing of change, so that there is an overall sense of "ownership" for the change.
- The concerned project teams (relating to this transition) are normally in a "standby" mode to resolve any issues and bugs that were not detected during earlier testing. However, it is the responsibility of the Program Governance Board to ensure that this "stand-by" arrangement is not extended beyond what is reasonably needed (as the concerned project resources may need to be released from this transition and redeployed for subsequent projects, etc.).
- The transition step may call for "parallel runs" or invocation of partial rollbacks, as needed. Typically, it is noted that more issues are discovered during the first flush of change; this can be a testing time for all and needs to be managed appropriately.
- Once the Functional Managers are convinced that the transition has taken place successfully, the concerned Project Managers can be released, while keeping some support resources, as needed. The

concept of the "learning curve" is evident here, as the operational resources become increasingly familiar with changed procedures/ systems, etc., and start assimilating them as a part of BAU. The outcomes begin to be stabilized here.

- More concerns start emerging about the change, and the Functional Managers need to be prepared to counsel the operational departments more, as needed.

5.4 Post-transition Step

During this step, the operational departments have sustained outcomes. Depending on change preparedness and acceptance, a self-sustaining change process is now set into motion. There are few early adopters for change, and some would like to wait and see what happens. We will discuss these aspects more in the section on Change Management (Chapter 6), but the rate of adoption for change can vary with the type of organization and the nature of the change (whether the change is compliance oriented or any leeway is allowed for it).

Once the outcomes begin to stabilize, the benefits start to emerge. The Functional Managers can review the respective benefit cards and invoke benefit-measurement systems. It is also likely that the benefits are sometimes measured by unbiased third-party agencies or through user surveys (as in the case of an increase in market share or customer satisfaction), as commissioned by the Program Governance Board. It is quite likely that, in some cases, there is a time-lag between the outcome achievement and benefits realization, to ensure that the new practices have "become part of regular practice."

The access to "legacy systems" is cut off, so that the organization does not relapse into "old ways of working," citing trivial issues, etc. This is more commonly seen in IT-transformation programs, where users have reverted to "legacy" systems when network connectivity and bandwidth issues surfaced.

If the benefits measurements indicate that the extent of benefits realization is not as originally planned, it becomes the responsibility of the Functional Managers to escalate to senior management [especially to the Portfolio Progress Monitoring Group (PMG)]. This step ensures that the linkage between the program benefits and the achievement of the strategic objectives of the organization is sustained. In response to these escalations, top management can take one of the following actions:

- Commission a new project (or even a program) that will address mid-course corrections towards achieving the stated benefits.
- Scale down the benefits target (which can occur due to factors that have changed, such as an increase in competition, resulting in a reduction in market shares projected).
- In extreme cases, when it becomes apparent that the benefits are not being realized as expected, or they can only be achieved with infeasible additional costs, the program can be terminated.

In all these cases, the Functional/Business Area Manager also records the lessons and provides reverse feedback about the strategy implemented. If the benefits were not realized as expected, the feedback can relate to whether it was a failure from a program delivery perspective, or if the program itself was incorrectly formulated to begin with. This is useful feedback for the senior management in further refining the portfolio.

When the actual benefits are measured, the benefit cards are updated. It is quite likely that the program might have been moved to the next iteration, so the operations need to brace for the next wave of change.

Transition management can be a grey area, as multiple stakeholders with diverse interests are concurrently involved. If managed well, it can be a "win-win" relationship for all. Poorly managed transitions breed skepticism, which becomes a dampener for introducing further changes in the organization.

Summing up, openness for beneficial change and sustained top management commitment are the keys for successful transitions. Major transitions are risky propositions, involving mindset changes from people. We will also cover Transition Management from William Bridges' model in Chapter 6.

Chapter 6

Change Management and Stakeholder Engagement

6.1 Significance of Change Management and Stakeholder Engagement

No change occurs in a vacuum. The very connotation of change implies modification to an existing system—be it individuals, teams, or organizations. The swiftness and the extent of change can vary with the circumstances. Since change affects human beings the most, we also consider how to address the impact of change on our stakeholders.

As noted earlier, the impact of change can be analyzed from three perspectives—at an individual, team, or organizational level. There is considerable literature available on how change impacts individuals and teams—for example, *Making Sense of Change Management*, by Cameron and Green.[1] Herein, we focus on how change impacts organizations, as most of the change initiatives we consider, including portfolios, programs, and projects, are oriented towards changing organizations. We will, however, refer to specific aspects of individual and team change, as necessary.

6.2 How Change Gets Triggered Off?

In Chapter 1, we discussed the context of change and that triggers for change can be due to internal or external forces. In addition, change can be swift or emergent. Natural disasters, legal decisions, huge stock price fluctuations, etc., can cause swift changes. Emergent changes usually occur when the trigger for the change is well known, and the seeds for change emanate from different divisions of the organization. These could be due to a long-standing decline in share prices, a decrease in customer satisfaction ratings, or product defects, to list a few. Internal triggers for change could relate to, for example, low employee morale, lower productivity, and wastage/rework.

In emergent change, the need for change is noted by many employees working in various divisions, but it may not be articulated openly because of company policies or the culture of the organization. However, when the need for change reaches a "tipping point," so to speak, it comes out into the open and needs to be tackled by Senior Managers or other leaders. Changes in the government normally occur due to the expression of discontentment by the voters with the previous administration.

In response to these changes, organizations can effect modifications to the portfolio, which can spawn new projects and programs, or changes in direction for projects and programs in progress. In the chapter on Program Management (Chapter 3), we noted the different types of programs that can be executed, including "top-down/vision led," "bottom-up or emergent," or compliance driven. The response of the impacted stakeholders may be different for each of these types of programs.

6.3 Enabling Changes from Enterprise-Wide Transformation Initiatives—Stakeholder Classification

Stakeholders remain at the core of the change impact. These are people or groups who are either:

a. Impacting the change (for instance, top management, governance and regulatory agencies, company shareholders, etc.)
b. Impacted by the change (for instance, line management, end users, and customers)

c. Those who are involved in implementing the change initiatives [Project and Program Managers, vendors, members of the Project Management Office (PjMO), etc.]

It is also quite likely that some of the stakeholders (or stakeholder groups) fall under two of the above categories—for instance, customers can impact the change by changing their buying preferences and can also be impacted by it (through the use of newly launched products).

Many companies usually use the CPIG classification of stakeholders, as under:

C: Customers—Those who generally are impacted by the change
P: Providers or Suppliers: Those who provide goods/services
I: Influencers: Those who could influence the change, including government, shareholders, etc.
G: Governance: Decision makers or regulators

It is apparent that many stakeholders can fall under multiple classifications, and the communication messages, channels, and formats could be different for different groups.

Identification and classification of stakeholders are not straightforward in many engagements, and especially for external consultants doing a client engagement, it can be pretty daunting, as they probably would not be aware, at least initially, of the internal power maps of the organizations. The following techniques can be ideally be used to identify and prioritize stakeholders:

a. Use of group workshops: Here the major stakeholders are identified first. A second-level meeting may be held with identified stakeholders to uncover more stakeholders. Existing checklists and mind maps can be used to identify these additional stakeholders.
b. A variant of this technique uses individual and group meetings. Initially, the major identified stakeholders make individual lists of stakeholders. Then, these stakeholders are paired in groups of twos, wherein both the participants in the group compare their lists and identify common stakeholders noted by both the members and the stakeholders identified by just one member and debate whether such stakeholders are really material for the change initiative. After deliberations, additional stakeholders are included in the common

list by mutual consent. Thereafter, the groups of four are formed to repeat the process, to extend the list of stakeholders.

An advantage of this technique is that consensus is built during the process of preparation of the stakeholder list, and inputs from all of the core members are used. Some organizations use these techniques not only to identify but also to prioritize the stakeholders. Existing checklists from prior similar projects would enable these lists to be completed faster. However, these checklists need to reflect the realities of the current context. Intensive discussions during the group meetings, however, enable the Project and Program Managers to obtain greater insight into the interest and influence of various stakeholders and their classification.

In large organizations, as well as in Agile/Scrum project management contexts, stakeholders are represented by "persona," which is something similar to the depiction of their role. This description is typically more applicable when external consultants create a typical facet of a stakeholder in an empathy map. (This is a technique that is typically used in customer profiling.) For instance, Figure 6.1 shows an empathy map of a Manager of a Stores business unit in an organization. The organization has deployed an external consultant to trim its expenses, and one of consultant's recommendations is to reduce the personnel in each store. After interviewing a couple of Store Managers in the organization, the empathy map created may look similar to what is depicted in Figure 6.1 for a typical Store Manager whose nickname is "Jack."

As stated in the empathy map, Jack feels that he knows the sales patterns in his stores well, and that the Corporate Headquarters (HQ) or the deployed Consultant does not know actual political and ground realities well. He hears lots of "management jargon" all around about why the change should happen and how it will benefit the organization, and he senses a lot of uncertainty.

He openly says that the HQ does not know how the business runs in the stores, and he is open to change if it is proven to be beneficial to all. Routine messages on the project's progress from HQ are ignored, as he is not yet mentally attuned to the change.

Jack is pained by the possibility that he may lose some of his best salespeople if the staff reductions occur. He also recognizes that the change initiative is an enterprise-wide deployment and is not targeted only to his stores. He realizes that, ultimately, he may need to comply with the corporate directives.

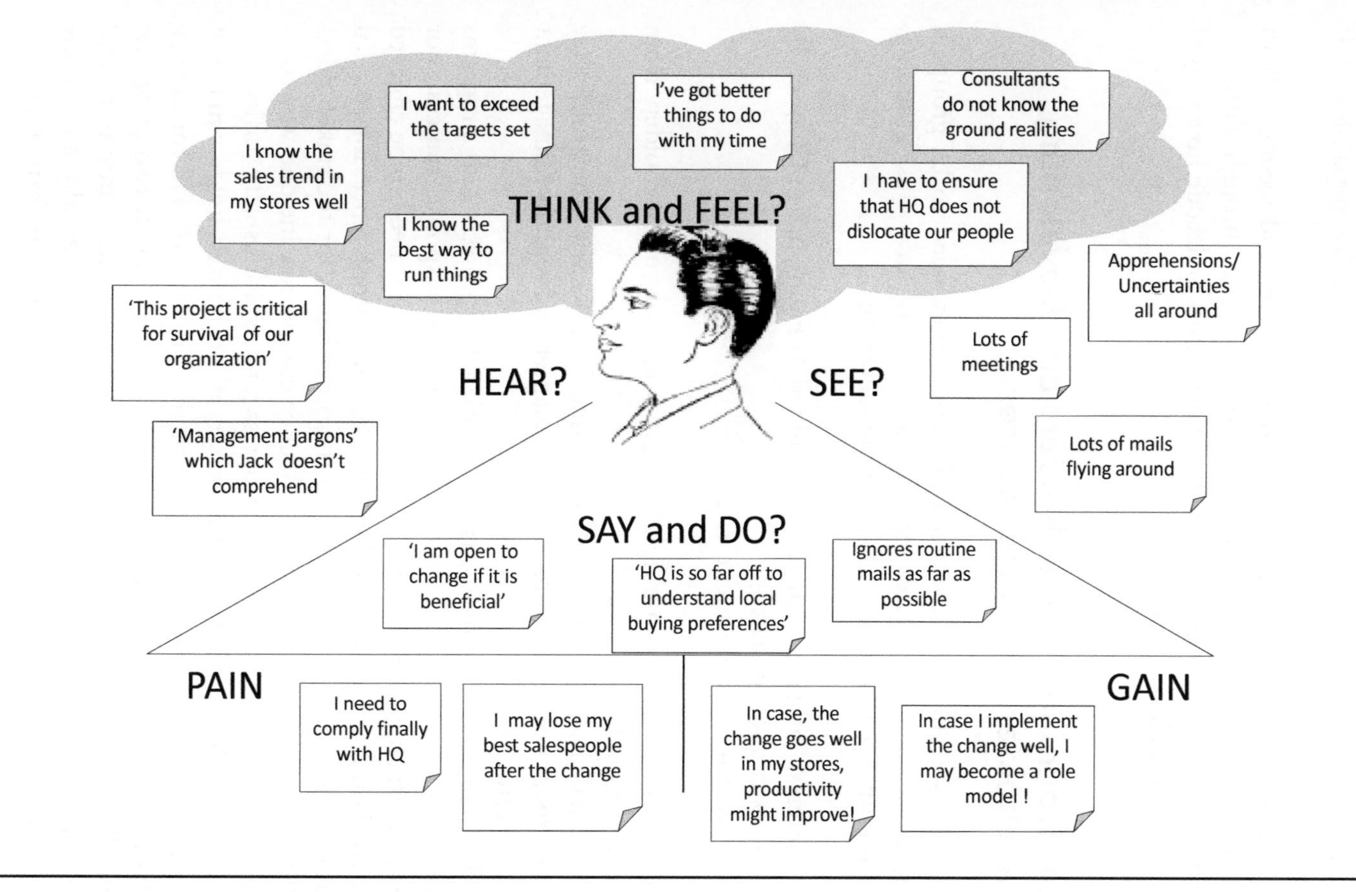

Figure 6.1 Empathy map for a stakeholder.

On the brighter side of it, Jack feels that, if the change does go well, he may be able to record higher productivity from his remaining staff people, and he may get a bonus, as well.

The empathy map portrays the mixed emotions and feelings—both stated and unstated— that typical stakeholders go through during a change. The empathy map is a powerful depiction to decode how to best address stakeholder concerns.

6.4 Grouping of Stakeholders and Developing Stakeholder Response Stances

After the stakeholders have been identified, they need to be grouped. Organizations use diverse techniques to carry out this grouping. Commonly used parameters include the following:

- Interest-influence matrix (as was noted in Figure 3.3 in Chapter 3).
- Impact and influence
- Power and commitment, etc.

It is preferable to have multiple groupings to analyze stakeholders from diverse perspectives.

The stakeholder response stance obviously varies with the group they belong to, as noted in Figure 3.3 of the chapter on Program Management (see Chapter 3).

The communications management plan for projects and programs should be closely linked with the stakeholder response stances. For instance, the high-interest and high-influence groups need to be engaged closely. Thus, periodic face-to-face meetings or teleconferences for a virtual team are needed for this group. On the other hand, for the stakeholders falling into low-interest and low-influence groups, routine one-way communications from the initiative's progress reports will usually suffice.

It should be noted that, at the outset, all of the stakeholders may not be enthusiastic about change. Some of the "innovators" fall for change instantly and, in turn, influence the "early adopters." These two categories typically account for about one sixth of the stakeholder community, and they value communication and propagate change internally among the rest of the stakeholders. After these two groups, come the "early majority" of stakeholders, who want to see initial successes to be convinced about the benefits of change. The early majority stakeholders usually constitute one third of all the stakeholders. The remaining 50% of the stakeholders fall into the category of "late majority" and "slackers." These groups would require considerable persuasion to adapt to change. However, any successful change initiative must penetrate into this last 50%, and once it reaches a "tipping point," then the change sustains itself.

6.5 Why It Is Difficult to Change Stakeholders? Or Stakeholder Views?

Although quite a few stakeholders understand the need for change, they resist or gloss over the impact because of personal biases or preconceptions. These biases are more applicable for the early majority of the stakeholders, who "wait and watch" for positive or negative signals to decide if they should embrace the change. Some illustrations of the cognitive biases that can hinder change are as follows:

- *Anchoring:* Relying on the first piece of available information to make a judgment about a change. This is especially significant if the early messages about the change initiative seem to indicate that it is not going well or is going too well. Many of the people tend to form a judgment based on these early trends, which may not be accurate. This is one reason that most of the programs tend to plan for early benefits to raise the image of the program to convince the early adopters.
- *Bandwagon effect:* When people tend to go with the collective view of others because they feel that the "majority is generally right." This bias operates predominantly during group discussions and brainstorming, when the "silent majority" goes with the viewpoints expressed by the "vocal few," as an effect of "group thinking."

  This bias is best addressed by the Project or the Program Manager by eliciting views from all concerned during group discussions. If the "silent majority" are not willing to come out in the open and speak because of cultural or similar factors, it is better to apply the "nominal group technique," which is widely used in idea generation in groups. As an application of this technique, people first work silently and independently write down plausible solutions to an issue and pass it to a "Scribe," who is more of a coordinator. The Scribe collects all of the ideas and ranks them. These ideas are discussed in groups before converging on a set of possibilities, with every group member having the opportunity to speak. This technique leads to an increased number of heterogeneous inputs, overall participation, and, subsequently, higher-quality decisions.

- *Automation bias:* Relying too much on automated information. This bias manifests with a few Project Managers, who tend to perceive whatever information and analysis is derived through integrated Project Management Information Systems (PMIS) as correct, without ascertaining if the underlying data-points were reliable and accurate to start with. Tools and techniques can facilitate (and many times support) and test the veracity of a hypothesis and the quality of decisions in real-life situations. This bias also impacts the organizational leaders, who tend to invest in excessive automation without cleaning up the processes or ensuring that the right work is done through the application of correct skill sets. Cultural change is a critical prerequisite in many organizations to ensure a higher quality of decision making, without solely depending on analysis based on automated systems.
- *Confirmation bias:* One of the more prevalent biases, where people tend to look for and absorb information that suits their preconceptions and ignore anything new. Many times this bias operates in tandem with the "anchoring bias," when people "discover" the first evidence of their beliefs in a situation that confirms their hypotheses, and they retain such impressions. This confirmation bias again manifests in many social situations and also impedes decisions in organizational behaviors by creating personal stereotypes. Supporting the judgments through analysis and data-points can address this bias to a large extent—provided the people who are having such biases are open to change.
- *Dunning–Kruger effect:* This cognitive bias has two dimensions. In some situations, people who are unskilled tend to overestimate their ability in learning new things or adjusting to change. Conversely, highly skilled people tend to underestimate the effort required by the relative newcomers to learn and adapt to change. In many situations, the top managers are already mentally attuned to change and are

looking forward to a new beginning, whereas the impacted Functional Managers are still coping with the change. This dichotomy creates disconnects and hampers the pace of the implementation of change initiatives. In a subsequent section, we will see the role of Change Agents and how they can facilitate bridging this "gap."

- *Planning fallacy:* Typical over-optimism concerning expected completion efforts of tasks or deliverables over which people feel they have complete control. This fallacy impacts the project planning, as some Project Managers tend to give optimistic estimates of schedules based on personal beliefs that they are in a position to control external events when they drift out of control. In many situations, the planning fallacy goes hand in hand with the optimism bias, whereby people tend to use the most favorable data for planning. Such a bias leads to the overestimation of benefits and boosting up the return on investment (ROI) during business case formulations. And an effect called "Pollyannism," or positive bias (where people tend to remember the most pleasant experiences during the decision-making process), is also strongly correlated with this bias.

  This bias can be addressed by subjecting the estimates to external scrutiny (such as through a PMO) and validating them through an independent assurance review process.
- *Self-serving bias:* One of the commonly seen biases—when the Senior Managers attribute all the successes to themselves and the failures to others and external factors. In change management, this situation is usually seen in "top-down" driven organizations, which tend to underestimate the role of other players (especially those people who are impacted by the change) and overestimate their own capabilities. A culture change is needed to address this bias toward promoting a collaborative working approach.

There are many more biases that tend to operate in the change environment, but we will not be discussing all of them here. It should also be noted that which biases affect a person depend on one's social and economic background and individual/organizational past history of adapting to change. Such biases, however, tend to slow down the change and create varying segments of people who adapt to change. Except in compliance-driven programs, change percolates slowly. Sustained top management commitment and repeated communication messages to the impacted stakeholders through appropriate media channels are keys to ensure success in such situations.

6.6 Applicability of Change Management Models in Driving Change Initiatives

There are several models related to change at the individual, team, and organization levels. Although developed independently of the project portfolio perspective, the applicability of many of the models is readily seen when dealing with change resulting during the implementation of change initiatives (especially programs).

A PricewaterhouseCoopers report notes that "there is an undeniable correlation between project performance, maturity level and change management. The majority of the best performing and most mature organizations always or frequently apply change management to their projects."[2]

We hereby describe a few of the change management models and how they can relate to the successful implementation of projects and programs.

A. *William Bridges' Transformational Change Model*[3]: As per Bridges, transition is a personal and psychological process of letting go of past patterns and engaging with new ones. It can be traumatic to some people, depending on the extent of the impact of change and how threatened people feel by the change.

There are three phases through which people go during a transition, which he calls, "Endings," a "Neutral zone," and "New beginnings." We consider these three phases through an illustration of a program that involves job restructuring and role redefinition.

During the "Endings" phase, people become disoriented when they know the change has been announced (or that change is inevitable and has to be implemented). If the change is intensely impacting the stakeholders, they tend to vent their grievances and are not mentally prepared to move on to embracing the change. Bridges suggests that the Project/Program or the Change Manager needs to keep communication channels open, listen to the personal issues of the stakeholders, and provide details about what is changing and what is not. Although the Manager can also describe why the change is occurring, people are usually not in the right frame of mind at this time to absorb this information in detail. Thus, the Manager needs to empathize with the stakeholders and "allow" them to grieve on their "losses," even though some of them could be imaginary. Generally, people tend to "cling" to old ways of working and the usual power structure, which needs to be gradually discouraged.

Once the stakeholders understand that change in inevitable, they enter the "Neutral zone." Again, this can be an uncomfortable zone for these stakeholders, as they have not fully let go of the past and have not completely embraced the change. It is quite likely that some of the stakeholders may try innovative methods to minimize the pain associated with the change, which needs to be encouraged by the Change Manager. Quite a few "hit and miss" solutions can occur here, especially as some of the early adopters tend to try out change and may have initial failures. Help desk and other support systems need to be in place here to see these people through the Neutral zone.

Once the inevitability of change is fully understood, people tend to move on to the "New beginnings" zone, where they become mentally "in tune" with changes. The Project or Program Manager needs to communicate the rationale behind the change, as well as the new roles people will play after the change. Early wins are celebrated that encourage more stakeholders to embrace the change. It is also quite likely that some stakeholders may stay in the Neutral zone, whereas the Senior Managers are already in the "New beginnings" zone, which creates friction. If the Senior Managers' messages about the change are consistent and the impacted stakeholders feel that the change is for the good, they usually will move to the "New beginnings," eventually. This can be a testing time for all concerned, as it takes time for the new processes to become familiar ones that are followed consistently. Previously noted biases against change can delay the transition process.

The pace of change also depends on the culture of the organization. Whereas in those companies with an Agile culture, changes can be quickly effected, it becomes more difficult in public sector/governmental types of organizations, where change tends to move slowly. If the change is "composite," involving multiple dimensions (including process, technology, etc.), it becomes even more traumatic for the stakeholders. Many organizational leaders then tend to stagger the changes, deferring the next change until the previous change has been fully absorbed by people in the organization, and the outcomes start to stabilize.

B. *John Kotter's Organizational Change Model*: Kotter[4] has written extensively on changes affecting entire organizations. He propounds that large-scale changes need to be managed through well thought-out and planned steps, rather than leaving the changes to evolve.

There are eight major steps in his organizational transformation model, which we will consider briefly.

1. *"Establishing a sense of urgency":* Top management needs to establish a sense of urgency about why the change is necessary. This can be done through an analysis of company performance, market forecasts, impending changes in regulations, etc. Kotter argues that a significant majority of the Senior Managers need to be convinced that the change must happen; otherwise, the impacted target stakeholders will not "unfreeze."
2. *"Creating a guiding coalition":* It is imperative that change in a large organization cannot be driven by a select few top managers. Rather, such top-driven changes breed skepticism, dampening the chances of success at the outset. In this context, Kotter suggests that a coalition of Senior Managers, cutting across Functional Managers, be formed to drive the change. This coalition can also consist of Change Agents from the Team Manager level, who can take on the role of connecting with various layers of the organization.
3. *"Developing a vision and a strategy":* Vision in a change environment portrays the end state after the change. The vision statement acts as a beacon for change, usually is brief, and should be able to be communicated to a wide variety of stakeholders to inspire them to a better future. The Sponsor of the change is usually vested with the responsibility of creating the vision statement and communicating it to senior management for buy-in and support. The vision statement is accompanied by a detailed strategy outlining how the company intends to achieve it. This strategy provides a conviction for the second-level managers of the importance of change as perceived by the Senior Managers.
4. *"Communicating the change vision":* This is essentially in a continuum with the earlier step, where the vision statement is developed in consultation with the stakeholders. In this step, the vision statement is propagated extensively throughout the organization. Typically, multiple and diverse channels are used to disseminate the vision,

as people tend to gauge the seriousness of change through repetitive, consistent messages from managers. It has also been noted that stakeholders tend to absorb the need for change if it comes from executives and their immediate managers. Therefore, the change vision needs to be prodigiously communicated at various levels by different layers of management, to step up the tempo for the change.

5. *"Empowering employees for broad-based action":* Once the change gains momentum, more stakeholders are brought into the fray. This is the time to gain early wins and remove the unhelpful organizational structures, processes, and procedures, etc. These changes are welcomed by the early adopters, who need to be encouraged and given more in-depth training for them to propagate change further. It becomes the responsibility of the Change Sponsor to confront obstructive stakeholders to ensure that the change process does not get derailed.
6. *"Generating short-term wins":* It is imperative to obtain early gains to lend credibility to the change program and encourage more stakeholders. Once the employees are empowered for broad-based action, they become the catalysts for generating these early wins.
7. *"Consolidating change and producing more change":* Once the early wins are achieved, many organizational leaders tend to become complacent, which is a key reason for the failure of sustained change. The early wins need to be consolidated, and they need to reach a critical mass for the change to be sustained by people in the organization. This is a step wherein more and more stakeholders are brought under the ambit of change, and more Change Agents are needed to sustain the change. Executives can easily "take their eyes off," if they are not vigilant, as they may assume that the change will be self-sustained, when it may not be embraced as desired.
8. *"Anchoring new approaches in the culture":* Once the change is embedded in the system, it is self-sustaining and spreads across the organization. The initial set of Change Agents may get released (to anchor other changes), and a new set of Change Agents can replace them.

Kotter's entire approach indicates that sustaining change is not easy and requires protracted effort. There are a host of other models available for organizational change, and interested readers can refer to classic texts on change management—for example, *The Effective Change Manager's Handbook,* edited by Richard Smith et al.[5]

Edgar Schein[6] noted that there are two fundamental forces—one that impels the stakeholders toward change, and the other that resists change. The forces that induce change create survival anxiety in a person who could feel left behind when others have changed, or who could be viewed negatively by his manager if he or she does not adapt to the change. Peer pressure is another factor that induces change. The force that impedes change can be due to fear of the unknown, learning anxiety, or the familiarity of being in a "comfort zone." The survival anxiety must outweigh the learning anxiety before any change can happen.

Learning anxiety can be reduced by creating help desks and similar mechanisms that people can contact during the process of change. These aids are especially helpful during the "Neutral zone" phase described earlier in William Bridges' model. Financial and non-financial reinforcements will enable the stakeholders to move to "New beginnings" more quickly. Involving the learners and coaching models will also be helpful here.

For the change initiative to work successfully, it may be necessary to simultaneously reduce the learning anxiety and increase the survival anxiety. During mergers and acquisitions, for instance, each of the organizations has to learn about the cultures and specialisms of the others and also unlearn things that will not be relevant. The pace at which an individual accepts change can depend on multiple factors—one's personal history, the past history of the organization in coping with the changes, the nature of the change (if it is swift or emergent), and the type of the individual [as categorized by the famous MBTI® (*Myers*-Briggs Type Indicator®) personality types], etc.

6.7 The Ways in Which Different Organizations Work

Each organization has an inherent type of behavior, based on the industry it is in, cultural factors, and even its geographical location. Gareth Morgan[7] aptly describes the way organizations are structured around "metaphors." How change is initiated and driven depends, to a large extent, upon the metaphor that suits an organization. However, it should be noted that a single metaphor may not completely characterize an organization, which may adopt varying styles to implement different change initiatives. We next consider the major metaphors in which an organization can position itself and which style may be suited to drive change, based on the corresponding metaphor.

a. *Organizations as "machines":* This metaphor characterizes the organization as a collection of "cogs and wheels," which can be designed and controlled. Whereas this outlook may not be explicitly stated, the management's attitude and actions implicitly underpin this metaphor. Examples of such organizations could be government institutions, including the armed forces, and some the family-owned companies. Here, executives typically decide what needs to change (or deploys consultants to inform them what needs to be done). Detailed plans are prepared and executed and changes incorporated. Training is given to bridge the behavioral gaps, and resistance, if any, is handled by the executives. Whereas this metaphor may be ideal for changes of a compliance nature, such type of organizational leaders are often resistant to "listening" to feedback from customers to improve practices followed. Driving change becomes more of a "one-way street" in such organizations. Since reverse feedback loops are deficient, the organizational leaders discover too late that the change ought to have been managed differently, and another change initiative may follow to set the first one right!
b. *Organizations as "political systems":* In this metaphor, the organizational culture is driven more by groups and cliques. As compared to the formal power structure seen in the machine metaphor, informal power holds the sway in a political system. There are winners and losers, and people tend to rally around the winners to safeguard their interests. Allocation of limited resources is driven mostly by this informal power. Typical examples of this metaphor could include political parties, coalitions, and companies in which "favoritism" rules. In such organizations, work itself typically takes a backseat, and managers are more inclined to jockey themselves into positions of power. Change is driven more by political agendas and may take time to realize, depending on the continuity and commitment of the managers driving the change.
c. *Organizations as "organisms":* Organizations thriving in this metaphor receive feedback on their performance from their clients and try to improve. Such organizations are naturally receptive to triggers

of change and are open to innovate. Change is accomplished by congruence between individual work and organizational needs.

Thus, change in such organizations is driven through customer feedback and product reviews. Plans are set to achieve goals, and they are communicated to all concerned stakeholders. Reverse feedback is taken during the tollgates, and mid-course corrections are executed. Many of the modern organizations are generally classified under this metaphor.

d. *Organizations as "flux and transformation":* Change in these organizations emerges out of chaos. This is usually because the magnitude of the change is big, and a large number of stakeholders need to be convinced. One distinctive characteristic of this metaphor is that no single person or group is driving the change. Managers, at best, "nudge" the organizations to the desired end state, as well as by the forces of trial and error, and the organizations gradually move into the new equilibrium. This metaphor can be disconcerting for the managers, as they could feel they are not in charge of the change. But in reality, large-scale change initiatives tend to run this way, especially when limited direct "coercion" is exercised by top managers, and change proceeds on its own course. Managers can at best create the right conditions for the change to happen but not be directly driving the change.

The "flux and transformation" metaphor is also applicable when the need and rationale for the change is understood, but no single person or group is in a position to initiate or drive the change. This type of metaphor is seen more in societal changes in the case of burning issues. Discussion groups are formed, and gradually the momentum for change increases. Well-known techniques such as "World Café"* and "Open Space Technology"† become forums for like-minded people to gather together, debate, and initiate the process of change.

In "Open Space Technology" there is no "fixed agenda," and the participants experience the need to discuss a "pressing issue" that is at the top of everyone's agenda. This could relate to, for instance, social issues, such

* Co-originated by Juanita Brown and David Isaacs. (*Source:* http://www.taosinstitute.net/juanita-brown-phd#sthash.xNBU2fzM.dpuf)

† "Discovered" by Harrison Owens. (*Source:* https://en.wikipedia.org/wiki/Open_Space_Technology)

as "How do we address social inequalities?" This works when there are diverse opinions from the people who need to work together to discuss the issue in self-organizing meetings. "World Café" also relates to group discussions with large groups, where everyone contributes in multiple groups and the contributions are collated and shared. The above forums work well for changes driven by "flux and transformation."

For the change to be successful, different elements of the organization need to move together. This is especially true of the informal organization, which is not mapped to the formal organizational structure. A more holistic way of looking at change is to consider McKinsey's 7S model,[8] which takes into account seven interconnected factors that impact the change. These factors include:

Strategy: The way an organization builds and maintains competitive advantage
Structure: The way the organization is structured and the reporting lines are designed
Systems: The processes that govern the work
Shared Values: The core values and the culture
Style: The style of leadership adopted, including centralized or decentralized controls, and the quality of the leadership approach
Staff: The extent of staff deployed for various roles
Skills: The competencies of the staff working for the organization

Source: Adapted from Waterman, R. H., Jr., Peters, T. J., and Phillips, J. R. (1980). Structure is not organisation. *Business Horizons* Vol. 23, No. 3 (June 1980): 14–26.

All of the above factors need to be aligned, move synchronously, and reinforce each other for the change to be effective.

The bottom line is that how an organization is positioned (or finds a natural affinity with a metaphor) can influence how change initiatives (including projects, programs, and portfolios) are implemented.

6.8 Change Management Roles

Change cannot be brought about only by the Senior Managers. In order for the change initiatives to be successful, several roles are required to be in action. In the following, we discuss three major roles that can facilitate the change:

a. *Change Initiative Champion/Sponsor:* This could be the role of the person who commissioned the projects or programs, equivalent to the Program or Project Sponsor, and who is accountable for the success of the change. The Sponsor could also fund the change initiatives, authorize use of resources from other functions of the organization, and have the responsibility for creating and propagating the vision statement of the change initiative. It becomes the responsibility of the Sponsor to determine and discuss the benefits of change with other Senior Managers, convince them of the need for change, and confront any other Senior Managers who may be resistant to the change. The Sponsor has to act as the role model for the change and is an early adopter of the change. The Sponsor becomes responsible for communicating the change to the Line Managers, coaching them, as needed. The Sponsor is also responsible for designing roles and rewards and providing recognition to support change. From the portfolio management perspective, it is also the responsibility of the Sponsor to appropriately position each of the change initiatives with ongoing work and other projects/programs already underway within the organization.
b. *Line Managers:* These are the heads of the business units impacted by the change. They also need to sell the change to the Targets (those who actually deploy the change at the grassroots level, which could include the front-line managers). If the change initiative does not appear to be convincing, the Line Managers could have mixed emotions concerning the change. On the one hand, they need to inform the Targets about the change, and on the other hand, Line Managers need to be aligned with the senior management decisions (mainly coming through the Sponsor). In case the change initiative is not successful, this role bears the brunt of disgruntled Targets and cynicism when the next change initiative is launched.

   The chapter on Program Management (Chapter 3) notes that during the transition, the Line Managers need to ensure that there is no disruption of services to internal or external Clients.
c. *Change Agent:* This is the most significant role to enable change. The Change Agents can come from any rank, but, typically, the Team Leaders with a flair for communication are the ones most suited for

this role. It should be noted that Change Agents usually do not have formal line authority over the Targets (as Line Managers have). But the essential role of the Change Agent is to build connections, discuss the implications of change with the Targets, communicate the details of change, and give reverse feedback to the Change Sponsor and the Line Managers, as needed. Thus, the ability to work with a wide range of stakeholders is a critical skill for the Change Agents.

Change Agents also perceive additional opportunities during the implementation of change initiatives (especially for realizing supplementary benefits that were not foreseen earlier) and communicate them to the Line Managers and to the Change Sponsor. Change Managers or Agents can be internal to the organization (where the change initiatives get implemented) or may be people from external sources, typically from a Consulting Organization.

The advantage of an internal Change Agent is that he or she knows the organization and its culture well. However, the internal Change Agent may not be taken "seriously" by the top managers. In addition, there is pressure for these Change Agents to focus on their "day jobs," while simultaneously addressing the change requirements.

External consultants bring in wide perspectives and are generally perceived to be more unbiased, but it takes time for them to understand the nuances of the organization and its culture.

The involvement of Change Agents can also vary with the type of change initiative being planned. In compliance-driven initiatives (or "must-do" engagements), the change can be driven more easily by "telling" the target audience what they need to do and instituting rewards and penalties to induce change. For top-down engagements, management needs to inform the stakeholders about the need for change and the consequences if they do not embrace it. The "Machine metaphor" is more applicable in this context.

The greater difficulty lies with changes in which the target stakeholders are given the leeway to think through why they need to change and when they need to change. Typically, this context is applicable for the "flux and transformation" metaphor, but it can also be used in the "Organisms" metaphor. Though time consuming, the upside of this approach is that the stakeholders remain committed to the change. The role of the Change Agents becomes even more pronounced in such cases.

As we noted earlier, once the change process sets in, it can be reinforced positively or negatively. Positive reinforcement initially occurs when the target audience is skeptical but open to exploring change (as in the "Neutral zone" of William Bridges' model).

If the Change Agents continue to propagate change with support from top managers, early adopters become role models for others to follow. Negative reinforcement typically occurs when the top managers lose interest after the initial push for the change. When the Sponsor and the Line Managers "do not walk the talk," everyone gradually loses interest and the change is forgotten. This typically happens with internal change initiatives, such as in the creation of "quality circles." Creation of a Change Agent Network of committed people can address the negative reinforcement cycle.

When teams of Change Agent Networks increase, the familiar Tuckman model of "Forming/Storming/Norming/Performing and Adjourning"[9] becomes applicable. In some cases, the Change Agent Networks are "self-managing," without explicit control, whereas in some other cases, the Change Sponsor must provide strong direction and control.

6.9 Summary

In most of the cases, the change initiatives are planned well, but execution fails on account of poorly managed change. Change Management is essentially a "soft skill," which complements the "hard skills" associated with project or program planning and execution. The latter are necessary, but not sufficient to ensure successful deployment. Especially for large-scale transformation programs, change management becomes the "linking pin" for smooth transition to the target operating model.

References

1. Cameron, E. and Green, M. (2012). *Making Sense of Change Management—A Complete Guide to the Models, Tools and Techniques of Organizational Change,* 3rd ed. Kogan Page.
2. PwC Global Project Management Survey. (2004). "Boosting Business Performance through Programme and Project Management." PricewaterhouseCoopers. Available from www.pwc.com.

3. Bridges, W. (2009). *Managing the Transitions: Making the Most of Change,* 3rd ed. Da Capo Press.
4. Kotter, J. P. (2012). *Leading Change.* Harvard Business Review Press.
5. Smith, R., King, D., Sidhu, R., and Skelsey, D. (2014). *The Effective Change Manager's Handbook: Essential Guidance to the Change Management Body of Knowledge*. Kogan Page.
6. Schein, E. H. (1985). *Organizational Culture and Leadership.* Jossey-Bass.
7. Morgan. G. (2006). *Images of Organization,* 2nd ed. Sage Publications, Inc.
8. Waterman, R.H., Jr., Peters, T. J., and Phillips, J. R. (1980). "Structure Is Not Organisation." *Business Horizons* Vol. 23, No. 3 (June 1980): 14–26.
9. Cameron, E. and Green, M. (2012). *Making Sense of Change Management—A Complete Guide to the Models, Tools and Techniques of Organizational Change,* 3rd ed. Kogan Page.

Chapter 7

Benefits Management—Link between Portfolio and Program Management

7.1 What Is Benefits Management?

In the chapter on Portfolio Management (Chapter 2), we considered how to set the strategic objectives of the organization. These strategic objectives are usually set at a high level, with overarching organizational goals. One of the leading organizations we worked with had set for itself the strategic objective of being among the top three players in the industry within the next five years. In order to achieve this objective, it had to identify the group of benefits it needed to target that would facilitate this objective.

Benefits are measurable improvements resulting from change that contribute to the achievement of strategic objectives. In the previous example, the organization that we referred to had identified a number of benefits, including the following two:

a. To increase the revenue share of its flagship products by 15% in five years
b. To expand its geographic coverage to increase the revenue share from underserved markets by 10%

In the chapter on Program Management (Chapter 3), we also covered the concept of the benefits map, with intermediate outcomes facilitating the achievement of the benefits.

Benefits management relates to the cycle of identification, quantification, planning and realization of the benefits. The core objective of program management is to achieve the benefits that are closely linked with strategic objectives. Thus, benefits management is seen as a "linking pin" in many organizations between program and portfolio management.

Benefits should be perceived as adding positive value to the organization by the stakeholders. Thus, stakeholder engagement, which we covered in Chapter 6, is also strongly connected with benefits management.

7.2 What Are the Practical Issues Concerning Benefits Management?

One of the major issues concerning effective benefits management is relatively long lead times between its facilitation and realization. This is more applicable to project management, as in many cases, the facilitating project could have been over well before the realization of benefits. Benefits management has a closer linkage to the program management lifecycle, as the very purpose of the programs is to achieve outcomes and benefits. Because of the long lead times, some of the initiatives can get shut down (or investments reduced) before the realization of the full set of intended benefits.

In some cases, the baseline data for benefits measurement is deficient or outdated. This could be more applicable to societal programs—such as targeting the percentage of people living below a specific income cut-off level, etc. If the base values of benefits are not measured correctly, it will not be possible to accurately quantify the extent of benefits realization.

A third issue concerning benefits management is the measurement of benefits that cannot be readily quantified—such as customer satisfaction, employee morale, product attractiveness to the market, etc. Proxy measures can be deployed in some of these situations; however, these may not be adequately measuring the benefits.

Benefits ownership is another issue concerning management. Typically, the Functional Heads (e.g., the head of Marketing and Sales, Chief Financial Officer, etc.) are the recipients of the benefits from the change initiatives. The benefit facilitators include the Project and Program Managers, Team Managers, Change Managers, and a host of other stakeholders. During the benefits management lifecycle, especially during benefits planning and transition management, close coordination is required from multiple stakeholders. However, once the project outputs have been transitioned to operations, concerned Project Managers could disengage, and the role of the Program Manager shifts to oversight of benefits management (and is not directly involved in benefits realization). The primary responsibility of Functional Heads is to run the business. Because of this preoccupation, they could take their "eyes off the benefits." In the chapter on Change Management (Chapter 6), we covered in detail how important it is to sustain the change to ensure realization of the benefits. This challenge is real for many companies in which short-term results get precedence over long-term growth.

As in project management, the real challenge for benefits management lies in the gap between "knowing and doing." Due to day-to-day pressures, organizations struggle to implement benefits management effectively, in spite of "knowing" what needs to be done.

The integration of benefits management needs to be done along with project portfolio management and performance monitoring (as benefits are typically measured in operational areas or the impacted areas).

7.3 Benefits Identification

The first major step is to identify the benefits. Some of the organizations conduct benefits identification workshops with representatives from Senior Management. The Portfolio Steering Group (PSG) referred to in the chapter on Portfolio Management (Chapter 2) could be an ideal group for a workshop. The managers present in the workshop should be empowered to launch change initiatives (especially the programs) to realize the identified benefits—or benefits identification could become a one-off exercise. A strong facilitator is needed to conduct this workshop. Some

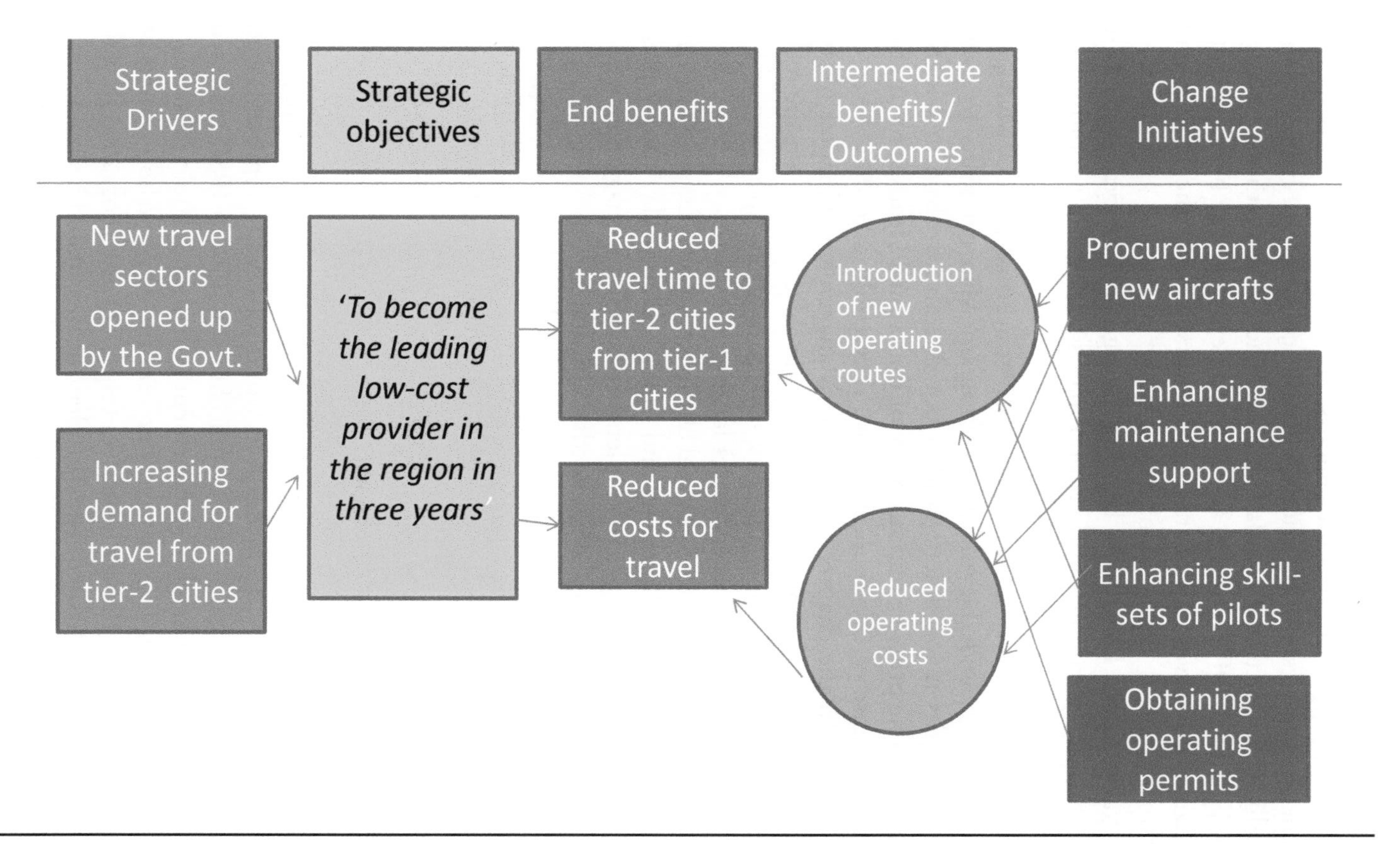

Figure 7.1 Benefits logic map.

of the larger companies create the role of a Benefits Realization Manager (or a similar designation) who is responsible for ensuring that benefits are identified correctly and that no double counting of benefits occurs, etc.

In organizations that have ongoing change initiatives (e.g., projects and programs), benefits can ex post facto be identified by conducting "benefits discovery" workshops. Ideally, this would have been when the change initiatives were first initiated, and the justification of the business case would have been based on the anticipated benefits. If this was not done appropriately or completely, the benefits discovery workshops formalize the benefits associated with each of the major change initiatives. Such discovery workshops enable the linking of the benefits to the outputs/outcomes of the change initiative. In addition, they build commitment among the key stakeholders for benefits realization.

These meetings can also produce the benefits map through discussions among the key stakeholders.

An interesting variation of the benefits map is the "benefits logic map." In this map, the strategic drivers (triggers) are plotted on the left-hand side leading to the strategic objectives (which could be goals). These objectives lead to the end benefits. Solutions/initiatives are designed to produce intermediate benefits/outcomes that lead to the end benefits. An illustration of the benefits logic map for a low-cost airline carrier organization is provided in Figure 7.1.

The advantage of this mapping is that the left-hand side (strategic drivers/strategic objectives and the end benefits) can remain relatively static, and updates can add to the initiatives and the intermediate benefits/outcomes on an ongoing basis. The role of portfolio management is apparent on the left-hand side of the benefits logic map. Projects and outcomes/intermediate benefits from the projects/programs link up independently from the right-hand side. One other variant can be added to explicitly state the assumptions underpinning the realization of outcomes and benefits.

The merits of producing the benefits logic map are as follows:

- Since the benefits-led approach is taken at the outset for initiative justification, a redundant or superfluous activity-oriented approach is minimized, and "pet or infeasible" initiatives are screened out during the initial justification itself.
- Assumptions linked to the realization of outcomes and benefits can be reviewed and validated on an ongoing basis. If these

assumptions become invalidated during the middle of a program or a project, the corresponding change initiative can be terminated, without wasting further efforts and resources of the organization.

- The benefits logic map (and the benefits map) highlights roles and responsibilities. The top management is primarily responsible for delineating the left-hand side of the benefits logic map (up to the end benefits). The Functional or the Business Change Managers are more responsible for monitoring the intermediate benefits. Finally, it becomes the responsibility of Program and Project Managers to define the initiatives to be taken to realize the outcomes. The linkages depict the joint interfaces of multiple roles.

7.4 Benefits Quantification

Once the benefits have been identified, we need to put in place the metrics to measure them. Financial benefits are usually amenable to precise quantification (as they may need to be aligned with standard accounting procedures, as per the financial perspective of the balanced scorecard). However, some of the non-financial benefits may require proxy metrics to measure them appropriately (as in employee morale, etc., as noted earlier).

Whereas finalization of the metrics to measure benefits may not pose huge challenges, issues can arise during the actual quantification and forecasting of benefits. One of the causes of not achieving the benefits is overstating the benefits—both current and predicted. Biases underpinning change management are also applicable in many cases here—especially the "optimism" bias—wherein people tend to take the most favorable data-points to project business growth, etc. Another bias, which especially affects benefits management, is the "sunk cost effect." An effect of this bias is that organizations continue to invest, even when the future benefits do not justify the costs required to realize them. However, since the organization had invested money in the past, future benefits are overstated to justify the continued investment.

The benefit forecasts need to be validated by the Portfolio and Program Management Offices, and such reviews need to deliberately look for disconfirming evidence against the benefits projections. Reference classes of comparable projects and programs are taken from historical databases during such reviews for assessing the distribution of outcomes and benefits and plotting the current initiative to assess the likelihood of its success.

Some organizations undertake an exercise called "pre-mortems." Here, during the launch of an initiative, it is supposed that the initiative has failed and the benefits have not materialized. The causes of the failure are "brainstormed," which may bring to the surface issues such as over-optimism, planning without adequate resourcing, and similar misplaced assumptions that went into the planning. Such an analysis enables mitigating optimism and similar biases even during the launch of the initiative and give an objective assessment of the likelihood of realizing the benefits. Probabilities can be assigned to various benefit values for better prediction. Techniques such as the Program Evaluation Review Technique (PERT) can be used to combine multiple estimates into a single value, thus reducing the risk of overstating or understating the benefits.

7.5 Planning to Obtain the Benefits

The benefits realization plan introduced in the chapter on Program Management (Chapter 3) incorporates the expected dates when the benefits are expected to accrue. It should be noted that in a program environment, "disbenefits" (which are perceived as negative outcomes) are also likely to occur. As an illustration, when the government funds the expansion of a city, more job opportunities are expected to open up for residents as benefits. It is also likely that environmental degradation occurs due to pollution, etc., which cannot be avoided in the context of this program. It is up to the Program Manager to reduce the extent of disbenefits and maximize the benefits.

Generally benefits can be "cashable" or "non-cashable." Cashable benefits typically occur when a company sells its nonproductive assets and recovers money. "Non-cashable" benefits can relate to, for example, "doing more with the same resources" or "the same work with fewer resources." Non-cashable benefits can also accrue due to higher productivity and better quality, etc. Some of the change initiatives are taken up to meet regulatory requirements. Such initiatives may not produce direct benefits but enable avoiding penalties, etc., in case of violations.

Benefits can be prioritized using "pair-wise" comparisons, which were described in the chapter on Portfolio Management (Chapter 2).

Apart from the benefits map, three other documents that are prepared during the benefits management cycle include the Benefits Management Strategy, the benefits realization plan, and the benefit cards. We saw some of these documents in the chapter on Program Management. (Chapter 3). In brief, the benefits management strategy is a guidance document that defines how benefits will be defined and managed throughout the benefits lifecycle. Benefit cards record each benefit and its categorization, details on how it will be measured, current levels, anticipated trajectory, risks, dependencies on other initiatives, and the details of the benefit owner. The benefits realization plan is a summary plan showing the consolidated view of when and how each benefit will be realized. The benefits map is usually included along with the benefits realization plan. These artifacts are usually more associated at the program level. The business case justification for the programs always involves consideration of the benefits, the likely costs of their achievement, and the return of investment thereof.

7.6 Realizing, Tracking, and Sustaining the Benefits

Benefits realization is mostly accomplished through the execution of programs, although stand-alone projects as a portfolio can also facilitate this. Benefits management interfaces with transition management, stakeholder engagement, and change management, all of which were discussed in detail during the previous chapters.

During benefits realization, the focus shifts to "making things happen." A sense of optimism and action needs to be pervasive here. The company's performance management system needs to portray the extent of benefits realized. It becomes the responsibility of the Functional Heads to monitor the benefits, assess if the actual benefits realization is deviating from the forecasts, and take corrective steps.

Too often, companies and businesses fall into the fallacy of creating capabilities and expecting results to follow without management interventions or mid-course corrections. Understanding what works and incorporating it into the system is a vital skill here. The benefits map produced earlier can be a useful tool for benefits reporting. Once the initiatives facilitating a benefit have been successfully completed, the corresponding initiatives can be code colored as green, and the focus can then move on to obtaining the corresponding outcomes. Transition Management reports (covered in Chapter 5) can give early warning signals if benefit realization is going to be hampered.

Mid-initiative reviews focus on whether the planned benefits can be realized as per plans, understand the causes of variations, understand if the disbenefits and "spin-off" benefits are appropriately addressed, if the updated business case continues to remain viable, and whether the governance concerning the benefits management itself (especially the adherence to benefits management strategy guidelines) is being followed appropriately.

There are a couple of techniques that aid in benefits reporting (especially for non-financial benefits). The most fundamental technique concerns the authenticity of the values of benefits reported through "one version of truth." This technique is concerned with benefits values being derived from an agreed-upon source and reported as per the agreed-upon schedule in the benefits realization plan. As an illustration, the perception of a new product launched in customers' minds can be measured through multiple market surveys. Where these surveys give divergent results, management needs to decide which survey result (or combination of survey results) is the most authoritative to interpret and take corrective actions and stick to the source consistently.

Whenever benefit cards are prepared, usually the "tolerance" limits for benefits realization are set. For example, if an automotive manufacturer expects to increase its market share for its brand from the current level of, say, 10% to 15% over a course of three years, it may also set a "tolerance limit" of, say, 0.5%. In other words, if the market share after three years is expected to be in the range of 14.5% to 15.5%, the change initiatives to achieve this benefit can continue. If the forecasted market share is expected to deviate outside the above limits, then mid-course corrections are expected to be launched. The tolerance limits indicate that it is not always possible to reach 100% of the value of the benefits, as planned. Due to unforeseen circumstances, there could always be deviations. How much of the tolerance the management is willing to accept, depends on the risk appetite of the company. The concept of setting tolerances and escalating only in the event of an exception conserves management time and efforts in tracking the benefits.

The major steps in benefits realization and tracking (after benefits identification) thus include:

1. Validate the baseline values of benefits. This is especially required when the baseline values were measured during the commencement

of the program and conditions have changed to warrant their re-baselining.
2. Participate with the corresponding project teams in providing requirements, and monitor the project progress. It has been noted that the very act of rigorous project monitoring enhances their chances of success.
3. Prepare the impacted operational areas for change.
4. Perform quality control on the project outputs (relating to user-acceptance testing).
5. Monitor transitions and report outcomes realization (see Chapter 5). It may be necessary to align rewards with outcomes and benefits realization to motivate the concerned stakeholders. The forward "booking of benefits" also facilitates an implicit commitment to achieve them.
6. Track and report benefits.

It needs to be reckoned that a focus on target setting and monitoring in isolation may not be conducive to enhance performance. In organizations that are obsessive about the achievement of targets, sometimes the benefits values are "dressed up" and target achievement can come at the expense of Corporate ethics. Case studies concerning the folding up of companies such as Enron, Worldcom, Lehman Brothers, etc., bear ample testimony to this.

Benefit reviews span the entire benefit lifecycle, which can extend beyond the project and program lifecycle. These include reviews of primary benefits that were planned, "spin-off" benefits that were not planned but emerge because of the realization of primary benefits, dis-benefits, as well as the review of governance mechanisms around benefits management. During the program lifecycle, these reviews can coincide with the "phase/gate" reviews of the initiative. In publicly funded initiatives, these reviews can also be undertaken by independent auditors or legislative committees.

Lessons learned during the benefits review are factored back. This process is more effective when the organization has a centralized Portfolio Office (PfO).

In large organizations, the formal benefits management process is ingrained within the Portfolio Benefits Management Framework, which could be owned by the PSG. The portfolio benefits management framework needs to contain guidelines for post-implementation reviews. Normally, such reviews are commissioned by the concerned Functional Managers and can be undertaken by the internal or external audit teams. The queries that are asked in such reviews can include the following:

- Were the forecasted benefits realized? If not, what is the extent of shortfall and the reasons thereof?
- Did the initiative represent value for money? If not, why? This analysis needs to be correlated against the information available at the time of investment.
- Were the good practices concerning benefits management appropriately followed, and what can be learned to improve delivery and benefits realization in the future?

A comparison of expected values versus actual values of benefits realized needs to be added to the database of projects for future estimation purposes.

7.7 Benefits Management from the Portfolio Management Perspective

As noted earlier, at the portfolio level, the benefits management process gets integrated. In such organizations, the PSG (or a similar body) ensures that a consistent benefits management approach is followed for all initiatives. During initiative selection, the PSG ensures that change initiatives selected represent value for money, and duplication of benefits is minimized. Normally, a benefits management framework document is created at the organization level for guidance. This document contains details on the following:

- How are the benefits identified and categorized?
- What type of standard measures and indicators can be used to assess benefits realization?
- How can benefits management be integrated with change initiative planning (including business case validation, etc.)?
- How can the organization ensure that benefits realization is consistent with its performance goals?

- How can the organization ensure that appropriate arrangements exist for benefits tracking and aggregate reporting at the portfolio level? (This can include the formats of the management dashboards for monitoring the benefits.)
- How are risks assessed concerning the benefits?
- What are the procedures for conducting post-implementation reviews and updates to lessons learned?
- What are the roles and responsibilities for benefits management?

Executing change initiatives in a portfolio environment leads back to the concepts discussed in the chapter on Portfolio Management (Chapter 2). Organizations need to ensure that the operations are ongoing to keep the business running, and that the demands of other change initiatives "in-flight" are addressed as well. Senior Management can take the following approaches for portfolio implementation. How benefits accrue to the performing organization can depend on the approach preferred.

a. *"Big bang":* Here, the implementation of change initiatives is taken as a "top-down" approach with a time-bound implementation plan. This is more appropriate for higher maturity organizations and when the organizational environment is relatively stable.
b. *Phased:* In this approach, a staged approach is taken in the areas of greatest opportunity, or where there could be possibilities of quick benefits. The overall portfolio and the approach to benefits management keeps evolving, considering the lessons learned, etc.

   This approach is more suitable if a significant number of key stakeholders are yet to be convinced of the change, and the top management is keen in pushing the change. The "flux and transformation" metaphor referred to in the Change Management chapter (Chapter 6) could be a context for implementing this approach.
c. *"Ad hoc":* Here, there is little direct Senior Management support available for change. Some of the Functional Heads, who would like to initiate change, try them on their own and showcase the value of formal benefits and initiative management. There is no assurance that the change will proceed as planned, and benefits may be realized sporadically.

Obtaining a commitment and consensus from Senior Management for large-scale investments is always an issue. Many companies use the technique of "collective deciding and speaking with one voice" for change management. Individual managers are free to raise their views and objections in meetings convened to discuss which change initiatives are to be taken up and which benefits to target, etc. However, once a consensus is reached, the managers "speak" in a single voice to the impacted stakeholders, to drive change.

Chapter 8

Setting Up and Running an Enterprise Project Management Office (EPMO)

8.1 Why an EPMO?

Entities in private and non-profit/governmental sectors need to constantly address changes that are due to external and internal forces. As noted in the chapter on Portfolio Management (Chapter 2), these organizations maintain the master portfolio/sub-portfolios and launch programs and projects to address the change. As the extent of change becomes more complex, these organizations are confronted with the following queries:

- Is our current portfolio aligned with our Corporate strategy?
- Are we responsive enough to address the impacts of change?
- Did we select the right programs and projects to run? Are we investing in the right initiatives?

– Are we deploying our scarce resources appropriately to manage change? Are the change initiatives we run optimally balanced/sequenced to address our capability and capacity constraints?
– Are our governance processes and value tracking systems robust?
– How do we ensure consistency in delivery and enhance our delivery capability?
– Are we getting the right benefits from our change initiatives?

Enterprise PMOs (EPMOs)—covering portfolio, program, and project offices—can enable the addressing of all of the above questions with the goal of safeguarding the business value and enhancing the delivery competency of organizations.

8.2 What Is an EPMO?

An EPMO is a combination of permanent and temporary offices set up to support change initiatives at various layers in an enterprise. The EPMO can include a centralized permanent structure to support the portfolio management functions, together with a group of temporary offices to support specific programs and projects. Some large companies can have permanent multilayered "nodal or hub" Portfolio Offices to address regional and SBU-specific requirements. The structure of the EPMO, to a large extent, depends on the P3M (Portfolio, Program and Project Management) maturity of the organization and the number of change initiatives it runs.

8.3 What Would Be the Structure of a Full-Fledged EPMO?

A full-fledged EPMO would thus include the following offices supporting each other from multiple perspectives (please refer to Figure 8.1):

a. A centralized permanent Portfolio Office
b. "Nodal or hub" Portfolio Offices, which are permanent, where the nodes could be geographic/SBUs, formed to accommodate region-specific needs and sensitivities

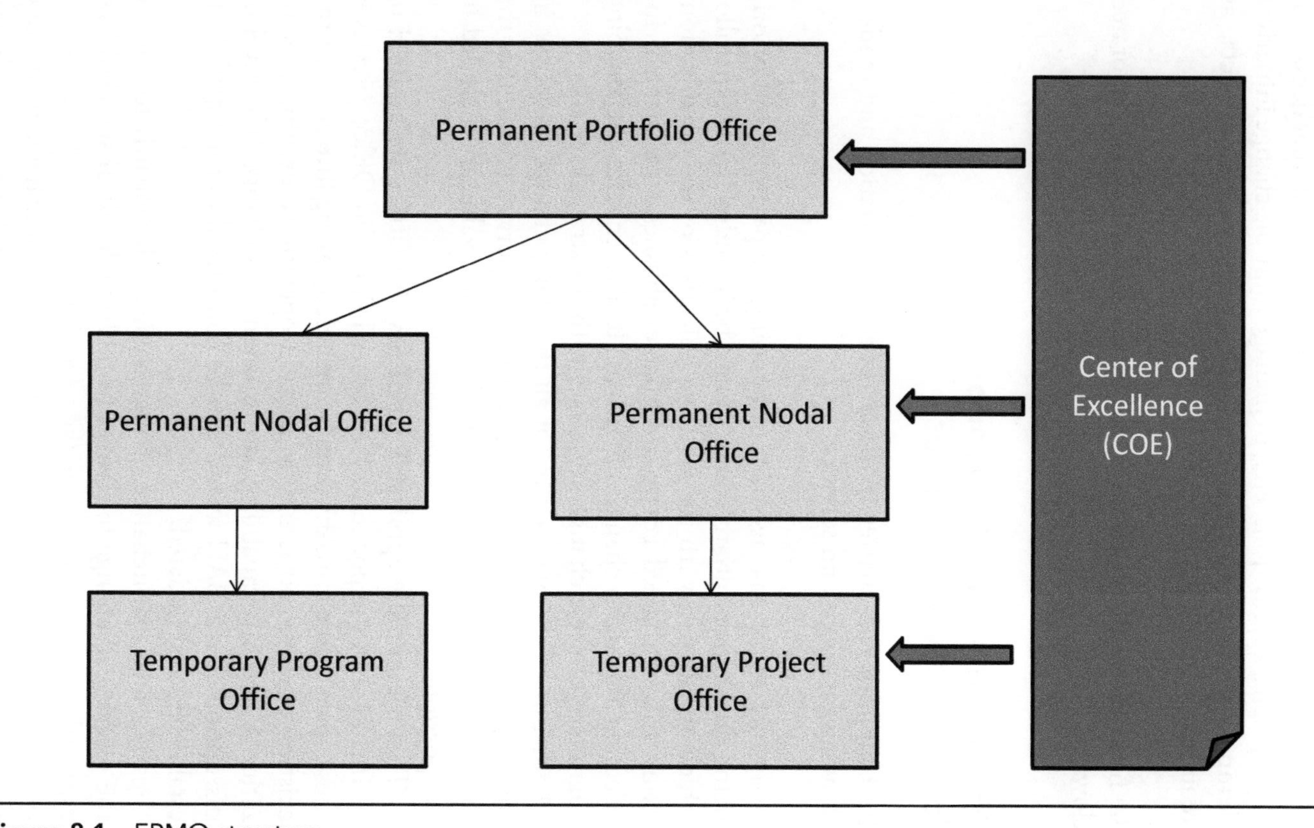

Figure 8.1 EPMO structure.

c. A standing permanent entity known as the Center of Excellence (COE), with the purpose of propagating the best practices, tool deployment, applications, etc., to manage multiple change initiatives
d. Temporary offices to support specific programs and projects, as required

The functions and services provided by various offices are elaborated on below.

8.4 The Centralized Portfolio Office (PfO)/Nodal Offices

The PfO supports the "master" portfolio of the organization and renders the following functions and services:

- Assistance to top management in building the portfolio and prioritizing the change initiatives that should be included in the portfolio, which are aligned with the strategic objectives of the organization. Thus, the centralized Portfolio Office establishes a mechanism for selecting the right change initiatives. It also ensures the ongoing alignment of these change initiatives with its strategic goals.
- Maintaining the "big picture" of all the major change initiatives in the organization and providing decision support to ensure that the right projects and programs that add business value are included in the portfolio.

  This is especially applicable when new initiatives are assessed to be included in the portfolio. The central Portfolio Office assesses if these new change initiatives can be included, taking into account the existing portfolio structure, impact on operations, existing organizational capacity, capability, and resistance to change. The level of disruption to the BAU would also be considered as the portfolio is established and updated.
- Creation and maintenance of management dashboards to convey the progress of change initiatives to multiple levels of stakeholders, especially to the top managers/key decision makers.
- Assisting in governance, especially for robust risk management, issue escalation management, quality assurance, and assistance during gate reviews.

- Assisting in the allocation of the right resources to various initiatives in the portfolio, taking into consideration the interdependencies across these initiatives.

As noted earlier, large companies may have individual nodal offices reporting to the centralized Portfolio Office. These nodal offices can lend support for the prioritization of local change initiatives, resource management, country/SBU-specific dashboard configuration management, dependency management of multiple change initiatives running in the node, etc.

Both the centralized and nodal offices are responsible for tracking project/program progress, ensuring that the business value for the money is obtained, and that benefits are realized. Another key function they perform is to enhance the visibility of the change initiative progress to the top managers, ensuring their better tracking, and improving organizational accountability/transparency.

The PfO can also maintain a flexible resource pool consisting of Project/Program Management Subject Matter Experts (SMEs) who can be seconded to functional departments to start and run the programs and projects effectively. This structure is generally seen for "internal" projects/programs executed for a performing organization.

8.5 Center of Excellence (COE) Functions and Services

As the name implies, this office sets standards for program and project management planning and execution (including providing guidance for processes, templates, and tools to be used). In addition, the COE can assist the organization by:

- Establishing the lessons learned/knowledge transfer databases and keeping them up to date with quality information, as well as enabling their organization-wide access
- Maintaining and updating the best practices and the metrics for regular reporting
- Project/Program Management Information System (PMIS) tools management to ensure that people can effectively deploy it and receive training in new versions, as required
- Facilitating program and project competency enhancement by providing training, mentoring, coaching, etc.

- Providing internal consultancy to projects and programs
- Facilitating internal assurance/health checks to programs and projects

Since the COE is a permanent structure, it "retains" the "organizational memory" and enables enhancement of PM maturity through regular assessments for suggested improvement.

8.6 Temporary Project/Program Offices

These offices are set up and run with the concomitant projects and programs and are, thus, temporary in nature. Such offices provide a range of services to the respective projects/programs, such as:

- Assisting in project/program planning and resourcing; tracking its progress
- Identifying dependencies and assisting in their management
- Customizing dashboards and tool configurations within the context of specific change initiatives
- Providing expert services, such as facilitation during project kick off, project reviews, and closure
- Assisting the Program/Project Manager in specialized functions, such as procurement decisions, vendor management, finance, stakeholder/communication, issue and risk management
- Establishing and managing a program benefits lifecycle

Typically, the centralized Portfolio Office and the COE report to the Portfolio Director (or similar) at the Corporate level, who, in turn, reports to the CEO of the organization. Nodal offices report to the respective heads of the SBU or the hub (such as the Regional Director). The temporary offices report to the respective Project/Program Managers while maintaining a communications link with the centralized Portfolio Office.

8.7 How Is the EPMO Set Up?

Setting up the permanent office can be done following a program lifecycle. The following are the major steps envisaged for this setup, which can be customized to specific situations:

A. Understand the "As-Is" Status:

Before embarking on an EPM implementation, it is necessary to understand the "as-is" state of the organization. This can include assessing the status of current change initiatives of the organization, their alignment with Corporate strategy, resources deployed, issues faced, "pain areas," and the expectations of the top managers.

More importantly, the organizational maturity in terms of project and program management needs to be assessed. Some toolsets can be used, along with models such as the OPM3® (Organizational Project Management Maturity Model from PMI) and the P3M3® (Portfolio, Programme and Project Management Maturity Model), which is a trademark of AXELOS Limited, to facilitate this assessment.

If a current EPMO exists, the functions and services provided by the EPMO need to be assessed to see if they are meeting the executives' expectations. An outline goal statement for the proposed implementation (e.g., a vision for implementation) is created here as a communication mechanism to various stakeholders. The implementation team works with the Sponsor for the EPMO implementation to attain agreement with this vision statement. The implementation team needs to have multifarious skills, including project and program management, organizational change management, PMIS tool selection and configuration/deployment, and management reporting.

B. Define the "To-Be" Envisaged State After EPMO Implementation:

The implementation team needs to assess the following:

1. Who are the major stakeholders that are impacting/being impacted by the EPMO implementation, and what and their interest and influence levels?
2. What will be the target operating model for the EPMO implementation? This can be assessed by agreeing on the functions and services to be provided by the EPMO and which maturity level for P3M implementation the organization should strive to attain after the EPMO program implementation.

A brief overview of the P3M maturity levels is given towards the end of this chapter.

It is necessary to analyze the EPMO target operating model from different dimensions, as follows:

1. Which new processes for the change initiatives lifecycle management need to be in place? Usually, organizations have an existing business change lifecycle that can be adapted for multiple divisions/portfolio segments, as needed, for change initiative management.
2. What type of skill sets should the P3M implementation teams possess?
3. What type of P3M reporting structures need to be designed?
4. What tool support is required to support the P3M implementation/change initiative management?
5. What types of information flows, management dashboards, escalation paths, etc., will be supported for P3M implementation? Usually these dashboards can be extended to actual change initiatives when they get implemented.
6. How will the change initiatives be governed? What types of governance structures need to be in place?

These can be aligned with the organizational governance structure as needed.

The EPMO implementation needs to run initiatives to realize the target operating model, as above. Once it goes "live," the redesigned processes and tools can be used by various change initiatives for their planning and execution as a standard guidance.

C. Develop Metrics to Assess the Success of EPMO Implementation:

These can include, for example, reduction in the extent of cost/time overruns for projects/programs, extent of benefits realization against plans, compliance to processes (as assessed during audit reviews), efficacy of risk responses, stakeholder satisfaction, and maturity improvements in PPM implementation.

D. Identify Major Risks Concerning the EPMO Implementation:

All EPMO implementations do not go smoothly. The challenges being faced by the EPMOs can be multifaceted, depending on organizational maturity, and, more importantly, by the culture and top management commitment to this change. The implementation team needs to consider the following major risks:

- Top managers may not give sustained support for the EPMO implementation.
- EPMOs are seen as cost centers and are disbanded during financial crunches.
- Resistance to change by the impacted staff (even Program and Project Managers) may derail the implementation.
- EPMOs may lack a holistic view—too much focus on tools, processes, or dashboards—leading to a lopsided implementation.
- Lack of authority for the EPMO, so it gets relegated as an office for information management.

It is the responsibility of the implementation team to identify such risks and escalate them as needed to the EPMO implementation Sponsor (who should ideally be the Portfolio or Strategy Director) for appropriate resolution.

E. Develop a Business Case for EPMO Implementation:

The business case contains the major objectives of the EPMO implementation, proposed timelines and costs, major stakeholders and risks, and the value proposition. The Sponsor for the EPMO implementation is the owner of the business case, and it is his or her responsibility for selling change to top executives and securing the necessary funding for EPMO deployment.

F. Develop a Phased EPMO Program Implementation Plan:

This step considers the budgetary constraints, resourcing issues, and the need to enhance the overall P3M maturity to develop the right implementation plan. The EPMO Implementation Manager can prepare this plan in consultation with the Sponsor. The implementation plan and the business case need to be approved by the executive team before its deployment.

8.8 Run the EPMO Implementation Program and Its Closure

The EPMO implementation will include deployment of the right skill sets, communicating with the appropriate stakeholders, assessing the risks for EPMO implementation and addressing them, and periodic assessment of the progress.

Typically, this implementation can last from 12 to 18 months for large-sized organizations with several hundred change initiatives. It needs to be reckoned that in large organizations, the EPMOs need to be "refreshed" rather than implemented anew, in which case the implementation timelines can be shorter. Periodic reviews can be conducted to assess the progress and viability of the EPMO's business case.

Once the required outcomes have been achieved and the benefits stabilized, the EPMO implementation program can close, and the implementation team can be disbanded. The newly formed or augmented Portfolio Office and the COE can take over from here to render support to the portfolio and to other change initiatives on an ongoing basis. It is essential to collect necessary data-points and configure the PMIS appropriately before the sign-off occurs.

8.9 Setting Up and Running the PMOs for a Specific Change Initiative

Whereas the previous model focused on setting up and operationalizing a "permanent" EPMO (along with the COE), here the focus is on establishing a PMO function aligned with a specific change initiative (project or program). This PMO can be staffed from the resource pool provided by the central Portfolio Office, and it can customize the processes and templates provided by the COE specific to the context of the change initiative.

These temporary offices can render the following additional services to the specific change initiative concerned:

- Establishing the common tool support and information flows for the project or program to align with the reporting requirements of the top managers

- Maintaining the master databases and documents, including provision of configuration management services
- Assisting in the rapid start-up of projects and programs through facilitated workshops, assistance to the Project or Program Managers in risk and issue management (including those arising out of dependencies), stakeholder and communications management, etc.
- Assisting the Change Initiative Manager in updating the status and managing quality standards
- Assisting in specialized functions such as vendor management, budgetary control, resource management, and providing access to strategic and governance-oriented documents to concerned stakeholders specific to the concerned change initiative
- Conducting lessons-related meetings and providing reverse feedback to strategy and to the COE
- Providing benefits lifecycle management assistance for the programs and performance management
- Providing administrative and secretarial support to the Project/ Program Manager

8.10 Challenges Facing the EPMO and Their Possible Remedial Measures

The setting up and running up of EPMOs is fraught with many challenges, especially in organizations with low P3M maturity. The approaches to implementation can include the "Big bang," where the implementation is typically top-down; "Phased roll-out," based on geographies or divisions; and "Opportunistic," where the implementation essentially follows an "ad hoc" approach to start the EPMO in an organizational unit and showcase it later to the rest of the organization to achieve "buy-in." We consider herein key points to be addressed during EPMO setup and implementation.

- The executive commitment is crucial to set up and "incubate" the EPMO in its initial stages. The expectations from the EPMO become enhanced as time passes, and disenchantments can occur, especially if the early benefits are not visible to the top managers.

- The EPMO implementation and the deployment teams should include seasoned Project/Program Managers with the ability to win the confidence of other PMs. Sufficient authority should be vested in them so the Change Initiative Managers can provide status reports, as needed, and are able to inculcate the "good processes" and have confidence in the EPMO's services.
- The EPMO needs to be funded appropriately, as it undertakes both "development and governance" roles. The funding can come from Functional Managers, who obtain benefits from the projects/programs they undertake, whereas the funding for the provision of common functions/services/development of centralized standards can come from the Corporate shared resource pool.
- It is also important that the EPMO does not get into "turf wars" with Functional and/or Project Managers and become "entangled" in organizational politics. Project/Program Managers should not see PMOs as structures that increase bureaucracy, hinder progress, etc.
- As a ballpark figure, any organization that manages at least 20 change initiatives will most likely require a formal structure to plan, implement, and monitor such initiatives. Most of the large organizations have ongoing PMOs that sometimes provide disjointed services to various change initiatives. A consistent approach would enable better synergy and provide better services to the organization.

8.11 Enhancing the Organizational P3M Maturity

The P3M maturity at the organizational level addresses how effectively the organization defines and implements the portfolio management structure and the change initiatives. Most of the P3M maturity models follow the Software Engineering Institute's (SEI) Capability Maturity Model's (CMM®) five evolutionary maturity levels and customize them to the basic underlying frameworks for project/program and portfolio management. Additionally, the Organizational Project Management Maturity Model (OPM3®) from the PMI and the Portfolio, Programme and Project Management Maturity model (P3M3®) from AXELOS are other well-known models for assessing and enhancing the PM maturity of an organization.

Any maturity model assessment needs to provide answers to the following fundamental queries to the organization:

a. *"Where am I currently in terms of maturity?"*
This can also be ascertained through "self-assessment questionnaires" propagated by many industry providers and also through the OPM3® and the P3M3®. The responses to such questionnaires give an overview assessment of the organization's overall maturity state at various perspectives, such as the project, program, and portfolio.

b. *"Which are my strong and weak points?"*
These can be assessed through responses to the self-assessment questionnaire, perusal of project/program artifacts for existing change initiatives, interviews with management, and examining the governance systems. Every organization could be placed differently. For instance, some organizations could be strong in scheduling management but having poorly managed risk and escalation-management processes.

c. *"Which target level of maturity should I move into and when?"*
This is a critical decision to be made by the organizational leaders. Typically, organizations can move up one level at a time (as per broad levels of maturity defined below). Depending on the culture and acceptance of PM practices, the pace of the journey can be fast or slow. The level of investment and effort required may also vary depending on the scale and structure of the organization and how many change initiatives are ongoing at a given point of time. The organization's past track record in effective management of its change initiatives should also be considered.

d. *"How is the route map defined for enhancing the PM maturity and implementing it?"*
The EPMOs can play a critical role during the entire process of assessment of P3M maturity and enhancement. In this step, they can provide invaluable support to the project and program management community as well as the Senior Managers on what needs to be done to build on the strengths and reduce the weaknesses.

e. *"Which are the best practices against which I need to benchmark my organization?"*
The COE referred to earlier can render support with information on best management practices, that can be adopted for the organization.

f. *"How do I measure the progress?"*
Once the implementation journey to enhance the PM maturity commences, the organizational leaders will want to know where

they are, any mid-course corrections that need to be done, and the rationale for further investment. The COE, along with the finance function of the organization, can provide appropriate measures for consideration.

8.12 PM Maturity Model—An Overview

Most of the maturity improvement models follow the familiar "plan–do–check–act (PDCA)" cycle, with local variations. Likewise, the definition of maturity levels changes with the models. Herein, we propose a five-level P3M maturity model definition, which can be adapted by various organizations.

In addition, the interpretation of these levels can vary depending on whether we are assessing a project, a program, or portfolio management maturity in the P3M spectrum. For simplicity, we are describing the typical maturity-level characteristics at the project level. Similar definitions can be extended to the program and the portfolio levels with applicable variations.

Level 1: Awareness

The organization is aware that projects ought to be run differently as compared to operations or the BAU. Indeed, a few local projects managed by some functional departments can adopt a rudimentary project lifecycle. However, they are informal efforts without institutionalized progress tracking systems. In some cases, initiatives run by government or the local civic agencies are in this category.

Level 2: Localized

At this maturity level, a few local projects and a few "star" Project Managers manage their projects well. However, there is no uniform acceptance for structured project management methodologies at the organization level. This is why we refer to it as a "localized" level of maturity, wherein the competencies exist at repeatable and isolated personal levels, rather than being widely exhibited or distributed at the organizational level.

Level 3: Centralized

This maturity level involves organizations with formal, documented, and defined project management standards, systems, and procedures. All projects will need to consider this centralized standard, but are free to flex it to suit the scale of the project.

It is from Level 3 onward that the EPMO can have a key impact, and any large-scale organization should aspire to reach this minimum level of maturity to ensure consistent and repeatable delivery of its change initiatives.

Regrettably, many of the organizations are able to reach Level 2, but falter while reaching Level 3. Typical characteristics of such organizations can include, for example, the existence of documented project management processes, but Project Managers bypass them and adopt their own procedures. Another illustration is the existence of a project-level risk register, but it is incomplete without a description of risk owners and their responsibilities. Sustained Senior Management commitment can facilitate movement to Level 3.

Level 4: Quantitative

Here the EPMO keeps quantitative measurements of project-related metrics, such as time and cost overruns, extent of benefits achieved, resources utilized and phase/gate review assessment results. Typically, this level is possible once the organization tracks all the ongoing initiatives rigorously and maintains a history of the "data-points." Running a quality management initiative is also critical here. This quality management initiative will facilitate recording and updating the data on the quantitative parameters, as required at this level of maturity.

Level 5: Continuous Process Improvement (CPI)

Level 5, the highest level of maturity, is reached when the organization proactively manages the processes and PMIS to ensure that the projects run optimally. CPI becomes a sustaining process in the organization as part of an evolutionary cycle.

It is typically noted that a few divisions in the organization (i.e., the IT and New Product Development divisions) tend to reach higher levels of maturity sooner because of the inherent nature of the initiatives they undertake. Reaching Level 5 can indeed be an arduous journey for many of the organizations, and retaining this level of maturity can pose greater challenges. A well-set mechanism for training, coaching of new Project Managers, management commitment, and sustained investment in knowledge management are essential to maintaining the CPI level of maturity. Indeed, many mature organizations have Communities of Project and Program Management Practices to facilitate their ongoing maturity enhancement.

Chapter 9

An Integrated Case Study—Application of Project Portfolio Management

In this chapter, we bring together the salient points discussed so far as applied to a case study. This case study describes a scenario of a diversified conglomerate having multiple Strategic Business Units (SBUs) facing diverse business challenges. How project portfolio management enabled the conglomerate to improve its business performance and move towards profitability and excellence is the underlying theme of this chapter.

The core of the case study is based on a real-life situation encountered by the author as part of one of the engagements. Supplementary information has been added to relate to situations faced by many companies, to make the case more comprehensive. Client names have been disguised to maintain confidentiality.

9.1 Background: The Company—AXN Corporation

AXN Corporation was a leading conglomerate with diverse lines of business. AXN commenced its operations as an engineering company providing electronic-based controls for automobiles. It subsequently expanded as a software provider to telecom companies for their services. With surplus cash available from the earlier two lines of business, it also ventured into the non-banking financial institution business, specializing in micro-banking for providing seed capital for niche industries, and for servicing their financial requirements.

The three business lines (SBUs) were indeed diverse, but leveraging on the cyclic nature of respective business lines and changes in customer preferences enabled AXN Corporation to diversify and grow. (The concept of the BCG matrix discussed in Chapter 2 is applicable here.) The nature of these business lines was, however, quite different.

(a) The Electronics Control SBU business was tightly linked to the fortunes of the automotive Original Equipment Manufacturers (OEM), which in turn depended on the overall global economic situation and customer buying preferences.
(b) Software development for the telecom companies (Telecom SBU) was a highly profitable business, but it was quite stressful for the personnel involved because of the rapid changes in technology and demands for quick delivery. In addition, the applications developed had to adhere to the regulatory requirements of various countries for financial transactions.
(c) The Financial Services SBU of AXN Corporation was making good inroads into the market, which was highly subject to regulatory restrictions. The demand for this range of services was also tightly coupled with the fortunes of industries for which the lending was done.

9.2 Management Structure—AXN Corporation

The AXN group company was headed by the CEO, who was assisted by the Chief Financial Officer (CFO), the Senior Director of Strategy, and the heads of three SBUs, who were also designated as Senior Directors. Apart from this core group, the Human Resources (HR) Director and Corporate Services Director (who also oversaw the Information Technology and Administration services) were part of the senior management team at AXN.

Each of the SBUs had a similar structure for shared services, with their internal control of respective finance, administration, and business operations. Additional roles were co-opted as required. For instance, the Electronics Control SBU had a Research and Development (R&D) wing headed by a Director. And the Financial Services SBU had a specific position for Compliance and Risk Director. The Telecom SBU had strong representation from the marketing services. All the shared services in the SBUs were having a matrix line of reporting, both to the SBU Head as well as to the corresponding Services Head in the Corporate Office.

The Senior Director of Strategy in the Corporate Office was assisted by the Strategy Management Office (SMO), which initially provided the SBU Heads with basic information on how many change initiatives were operating in their SBUs and their status, as collected from respective initiative managers.

9.3 Triggers for Change

Because of the global economic recession and slower consumer spending, the customers of AXN Corporation were affected, which in turn impacted AXN's growth and profitability. AXN Corporation had missed its profit targets consistently for three quarters in a row. The Board of Directors (BOD) of AXN Corporation was highly concerned and asked its CEO to address the situation. It had also set a target of doubling the overall revenue of AXN Corporation and increasing the overall profitability from the current 15% to 25% during the next five years. The BOD also asked the CEO to cut down on superfluous expenses and trim down the headcount in AXN, if necessary.

Given the directive to double the revenue and increase profitability in five years, AXN top management went about this mandate systematically. As per the direction given by the CEO, the Senior Director of Strategy commissioned an internal study on the health of the business. Initial findings of this study indicated the following:

(a) AXN Corporation had lagged behind its competitors in its investments for new product development. This particularly impacted the growth of Electronics Control and Telecom SBUs.

(b) The operating procedures followed by the Financial Services SBU were obsolete, leading to considerable delays in approval for lending to customers.
(c) There seemed to be a general impression of overstaffing, especially in shared services, which could be reduced.
(d) The Information Technology (IT) applications supporting the SBUs were outdated. Different SBUs, based on the empowerment given to them, had invested in discrete IT systems, with different capabilities. A centralized IT system with distributed processing capabilities was lacking.
(e) Management and staff morale was generally low, as it was well known that the company was not making enough profit. Some of the highly skilled resources were seeking out other opportunities outside the company. The replacement costs of such skilled and experienced resources were fairly high.

9.4 How AXN Corporation Went About the Change?

The CEO of AXN Corporation agreed with many of the findings above and asked the Senior Director of Strategy to delineate the way forward to improve business performance. It was also agreed that systems and procedures required restructuring, and any top-level manager resisting change would be counseled initially, and, in extreme situations, would be replaced.

The Senior Director of Strategy decided to adopt a combination of "top-down" and "bottom-up" portfolio management approaches to bring about improvement driven by various change initiatives. The performance management statistics (such as the headcounts of various divisions, their operating expenses, etc.) also required close scrutiny.

As a part of its "top-down" approach, the senior management team of AXN went about identifying the direction they should be taking to achieve their goals. The Corporate vision, mission, and value systems were redefined to reflect the new business imperatives.

As a part of the Corporate mission statement, it was reiterated that AXN would continue to operate the three business lines it was in and develop them appropriately to meet the overall goals set by the BOD.

This was in keeping with AXN's core competencies and its brand value. AXN decided, therefore, to reinforce the capabilities of existing SBUs, rather than to diversify.

The quantified "vision" set by the BOD was taken as the stretch target, which AXN needed to achieve to reorient its strategy as well as operations. It was recognized by AXN executives that strategic change alone cannot result in targeted goals; it had to be backed up by superior operational performance. So a multipronged drive to address both change and operations was launched. Whereas more effective portfolio management is facilitated by tighter linkage to strategy and the application of balanced scorecards, improved operational performance and benefits management were to be enabled by the successful execution of change initiatives, including projects and programs, as a part of its portfolio.

9.5 "Top-Down" Change—How Did AXN Go About It?

A series of "external and internal" scans were initiated by the Senior Director of Strategy and aided by the SMO team to understand the competitive scenario and the internal situation. The first tool that was used in this context was PESTLE (Political/Economic/Social/Technical/Legal and Environmental) drivers for change, which was referenced in Chapter 1. By using this tool, the team analyzed the impact of external change triggers for the SBUs. Although all three SBUs were impacted by the PESTLE factors, the extent of the impact of the drivers was quite different across the SBUs. For instance, the impact of technological factors was far higher in Telecom SBU as compared to the Financial Services SBU (which was more influenced by regulatory factors).

Next, Porter's "five forces model" analysis was commissioned to understand the competitive scenario. This analysis, inter alia, brought the following factors into perspective:

(a) Threat of substitutes: In Electronics Control SBU business, inexpensive providers were flooding the market. In order to reduce costs, some of the OEMs were moving towards fitting less expensive controls. The quality of such controls was in question, which could adversely affect vehicle performance. The OEMs must be informed of any potential issues with vehicle performance resulting from the use of such inferior electronic controls. At the same time, the Electronics Control SBU could launch a new series of products with lower costs and acceptable quality levels to counter the inexpensive providers.

(b) The Telecom SBU customers were demanding lower prices from AXN; otherwise, they would switch over to other vendors. The marketing team of the Telecom SBU thus needed to consider closing deals with lower profit margins, in return for a guarantee of higher order volumes from its customers.

(c) The change in regulatory laws in some countries could threaten the profitability of the Financial Services SBU. More geographic diversification was needed here to grow the business.

The above analyses, together with extensive stakeholder engagement, led to the creation of the SWOT matrix. An extract of the SWOT matrix for this case is provided in Figure 9.1.

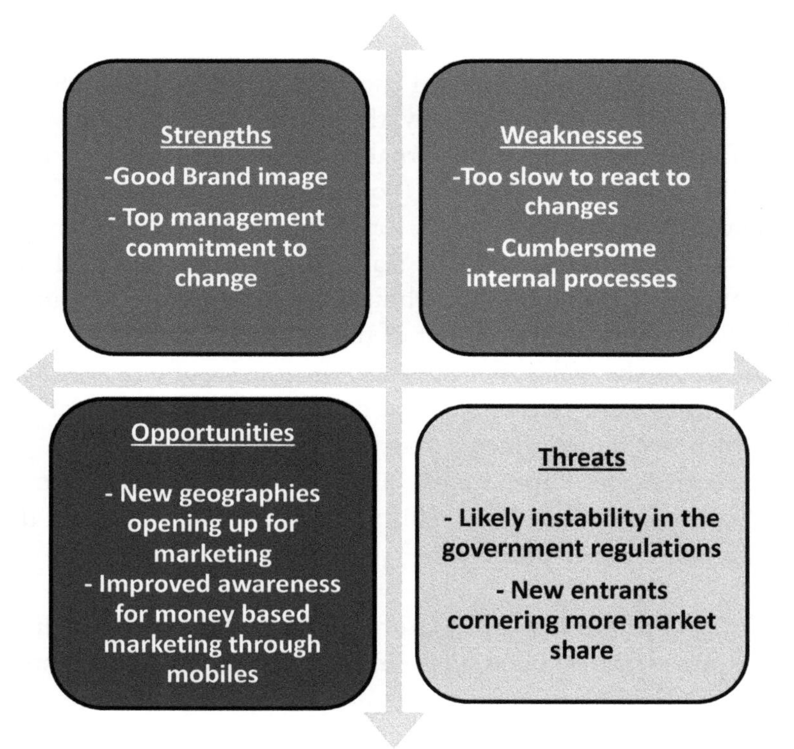

Figure 9.1 SWOT matrix.

Once the SWOT matrix was plotted out, risks and issues concerning the SWOT were analyzed. Again, this was done by the entire top management team through intensive workshops facilitated by the Strategy Management Office.

The top management team then decided which strategic direction to take for each of the SBUs. For instance, for the Electronics Control SBU, the focus was to add newer features to their products to differentiate them from the less expensive providers and then sell their products to the OEMs at a competitive price. For the Telecom SBU, the strategy chosen was to have a "customer intimacy/lock-in" by co-developing products suited to major telecom service provider customers. The Financial Services SBU was to use a "cost leadership" strategy to attract as many customers as possible. The target operating model and the growth strategies were thus quite different for each SBU.

For instance, the Electronics Control SBU had to put more focus on and investments in R&D and innovation-based projects, whereas the Telecom SBU's strategic initiatives were more focused on understanding the market and customer needs and providing bundled solutions at volume rates. The Financial Services SBU was more oriented toward rationalizing internal processes to reduce the cost of customer service management. Thus, the change initiatives taken up by different SBUs to reach their target operating models were diverse in nature. This phenomenon was in tune with the concept of "generic strategies" of Porter's "Competitive Strategy" model referred to in Chapter 2 (Portfolio Management).[1]

Thus, from AXN's perspective, several strategic themes were finalized from the balanced scorecard perspective.

(a) For the Financial Services SBU, the strategic theme selected was "provide the customers robust services with the least cost."
(b) The Telecom SBU selected the strategic theme "consolidate the customer base and get more repeat business by partnering with them."
(c) The Electronics Control SBU was more focused on the theme "introduce breakthrough products to gain more market share."

The strategic initiatives considered were in alignment with the above themes and were classified under the four balanced scorecard perspectives of financial, customer, internal process improvement, and learning and development.

As an illustration, the following strategic initiatives were considered for the Telecom SBU, which were in line with their selected strategic theme:

1. *Financial:* Sell off non-profitable assets to generate immediate release of cash. In addition, broaden the revenue mix across multiple products and enhance revenue targets from new products.
2. *Customer facing:* Segment the customers and undertake more effective value profiling to obtain more revenue from high net worth customers through product co-creation. Preferred customers were given access to premium products during their development, enabling the Telecom SBU to foster closer relationships with such customers.
3. *Internal processes:* Redesign internal processes to eliminate non–value-added processes and improve operational efficiencies. Establish exclusive call centers for the premium customers to provide quicker responses to their inquiries and for addressing their issues.
4. *Learning and development:* Enhance the ability of customer account managers to be more responsive and to create long-term relationships with premium customers. In addition, a separate initiative was launched to enhance employee satisfaction. Product awareness training for the new products was another component that was initiated.

The strategic themes represented outcomes that needed to be in place to realize the benefits and AXN's goals. These outcomes were to be achieved by the execution of inter-related projects towards creating capabilities within various balanced scorecard perspectives. In addition, the outcomes had associated performance metrics that needed to be measured after the transition to operations. The strategic initiatives were so designed to close the performance gaps for the strategic themes across the four balanced scorecard perspectives.

9.6 Reconciling with the "Bottom-Up" Approach

Whereas the previous approach enabled the creation of strategic initiatives from the "top-down" perspective, it was also necessary to reconcile the ongoing change initiatives (projects and programs) in AXN from a bottom-up approach. And here was the issue: The data on the status of many change initiatives were not centrally collated, as each individual SBU was running its own set of projects and programs. Some of these initiatives were consuming costly resources, competing with each other for critical resources, and were not aligned with redesigned strategic objectives.

The Senior Director of Strategy commissioned an enumeration of ongoing change initiatives through the SMO. This office was already receiving progress reports from a few critical initiatives, but many others were not under its radar. The SMO went about collecting key information on major change initiatives (above an investment cut-off determined by the Senior Director of Strategy). The information collected included the scope of each engagement, costs, intended benefits, resource utilization, and performance to date. This data collection exercise itself was a major challenge, as the SMO had to make a decision about which criteria to be applied while determining if a change initiative was better classified as a program or a large project. When these data were collated and mapped to the redesigned strategic objectives, it came as a shock to the AXN senior management.

About 25% of the ongoing initiatives did not map to any of the strategic objectives. These were mostly "pet" projects running under various SBUs. Many of the initiatives had duplicate outcomes. The top management of AXN saw a value in removing these duplicated, non-strategic initiatives. In addition, poorly performing initiatives were scrutinized in detail and terminated or merged with other initiatives, wherever possible. In this way, the top management improved the opportunity of deliverability of the overall portfolio.

The overall set of proposed new initiatives and the retained ongoing initiatives formed a first set for further work by the SMO.

9.7 Balancing and Deploying the Portfolio

Whereas the prior work identified which change initiatives were to be continued and, additionally, included in the portfolio, further work was needed to prioritize when each change initiative was to be executed and the types of investments required for the optimized set. The SMO served as the Portfolio Office (PfO) to undertake this work. The proposed change initiatives in the first instance were categorized under multiple "buckets" such as "Strategic," "High performing," "Support," etc. These categories were defined differently for the three SBUs. For instance, the Financial Services SBU had a separate category called "Mandatory" to address regulatory requirements.

The categorization was done for the initiatives under each SBU. In addition, initiatives supporting AXN's shared functions and the Corporate Office were considered during the categorization.

After this categorization, the SMO began to prioritize which change initiatives were to be executed as per different timelines. This prioritization was an involved exercise, as it had to balance the resource commitment for the implementation of existing change initiatives/BAU, quick wins required from new initiatives, achievability, and the attractiveness of various initiatives, to list a few.

We had considered the concepts of Net Present Value (NPV), Internal Rate of Return (IRR), and other financial metrics in the chapter on Portfolio Management (Chapter 2). These techniques were used by the SMO to prioritize the change initiatives. Priority was given to initiatives such as vendor selection, capacity building, selling off of assets to release cash, etc., which in turn could lead to the realization of early benefits or facilitate "downstream" initiatives to be taken. Wherever numerical quantification and prioritization were not feasible, alternate techniques such as multi-criteria analysis were used, and joint decisions were made with Senior Managers on the weighting and ranking criteria to be adopted and on the actual prioritization process itself.

The SMO's final step was to balance the initiatives as to when they should be launched, how the resource gaps would be addressed, and how to minimize the impact on operations. The updated versions of the Portfolio Governance Framework and the portfolio implementation plan (referred to in the Appendix) were guiding documents during the portfolio definition and its implementation.

Typically, it was noted that the Financial Services SBU had maximum ongoing interactions with customers. One of the objectives of balancing was to minimize the impact of the portfolio deployment on ongoing operations, which the Financial Services SBU specifically addressed.

The finalization of the portfolio was not a straightforward task for the Senior Director of Strategy and the SMO. They had to consult a wide range of stakeholders, take collective decisions from the top management, and ensure that mechanisms existed to support performance tracking of the portfolio. The findings were presented in graphical and other similar pictorial formats for easier understanding and assimilation by the Senior Managers.

There were several things AXN did right to define and institutionalize the portfolio. These included:

(a) There was a sustained top management commitment throughout the process. After the mandate from the Board of Directors, the CEO of AXN took upon the mandate as something that had to be achieved, and no "excuses" were acceptable. The CEO gave full power to the Senior Director of Strategy, who being a management board member, was driving the portfolio definition as its champion. This personal commitment not only enabled the portfolio to be defined appropriately but also smoothly executed, where many of the other companies faltered. The vision of the portfolio was in alignment with the Corporate vision, and it was communicated widely across all the SBUs and the shared functional departments.
(b) Reward and recognition systems were aligned. The performance targets were assigned to various heads of the SBUs, and the achievement of them was linked to the reward and recognition systems. The Key Result Areas (KRAs) of various personnel were defined in alignment with the achievement of the portfolio objectives, which were cascaded down to individual job role levels.
(c) AXN decided that the organizational governance and the portfolio management frameworks should be integrated. This approach was instrumental in avoiding conflicts about who escalates to whom and what the delegated authority levels are for decision making. Performance of the portfolio was reviewed in synchronization with the organizational performance review meetings, and a synergy was obtained. The SBU Heads were encouraged to monitor the change initiatives within their own jurisdiction through regular progress tracking and to take corrective actions as needed. These guidelines were noted as a part of the portfolio implementation plan.
(d) The application of balanced scorecard techniques was useful in portfolio definition, as everyone was clear as to which strategic initiative was contributing to which goal. Since the performance of these change initiatives was regularly monitored at various levels, a clear line of visibility was obtained on the link of the change initiatives to

the benefits. The benefits map at the program level was also a critical facilitating tool, driving portfolio investments.

(e) The SMO did commendable work in defining the portfolio and creating dashboards to report the initiative's progress as well as in establishing mechanisms for information flow. Without this support, the top management would have been confounded with the need for analyzing a whole mass of data to make decisions. Once the portfolio was established, the SMO (since it also doubled up as the Portfolio Office) was entrusted with the responsibility of tracking the portfolio and being involved in high-level progress reporting. Since the SMO was reporting to the Senior Director of Strategy, it was accorded a high organizational status, and the Program and Project Managers complied with the SMO's requests for progress reporting. Without this authority, the SMO would have been reduced to a coordinator of data with the need for it to "run around" and collect information from Project and Program Managers.
In order to facilitate progress reporting, the SMO together with the IT Department enabled procurement of an advanced PMIS tool. The functionalities of the PMIS tool were configured to match the business change lifecycle and the organizational governance workflows for AXN. This deployment enabled a seamless flow of data and collaborative work for decision making.

(f) We considered the need for appropriate change management for enterprise-wide transformations such as the one AXN executed. AXN top managers assumed that once the change initiatives were rolled out, the functional departments would, by default, be ready to absorb change. This did not happen by itself. Extensive communication events were subsequently planned and rolled out. There were "grievances" noted by the Managers whose initiatives had been terminated or reconfigured, which led to resistance In retrospect, more "listening" and engaging skills would have enabled a better "buy-in" from the impacted users.

(g) The skill sets of the SMO were important for buy-in from the Project and Program Managers (what we call the PPM community). Working collaboratively, the SMO was able to build relationships with key members of the PPM community, who in turn saw the value of the SMO provided.

(h) Once the portfolio was instituted, the Senior Managers recognized and supported the need for effective governance. As we noted in the chapter on Portfolio Management (Chapter 2), many organizations are adept at creating the strategy but falter during its execution. Obtaining the buy-in of AXN Senior Managers was a critical enabler for successful portfolio deployment. As a part of portfolio management, all the change initiatives (above different cut-off investment levels or those crossing a specific threshold) were monitored centrally by the top management of AXN. Within the SBUs, lesser priority initiatives were monitored by the respective SBU Heads.

Business analysts were engaged in refining the forecasts of benefits to ensure that monitoring groups were working with updated data. For all the change initiatives, the phase-gate reviews of the progress, including the continued justification of the business case, were done meticulously. Earlier, this rigor was missing, and many projects were allowed to "run freely," without clear justification.

(i) In many of the senior management reviews, benefits analysis formed the core agenda of discussions. This was especially true for the programs in the portfolio, as they were primarily focused on benefits management. Reliable benefits forecasting was a key to ensure that the discussions were meaningful. The SMO ensured this would occur by having the representation of specialists with benefits management experience in such meetings. In addition, in a large organization as AXN, it was quite easy to have multiple change initiatives delivering duplicate benefits. It was the role of benefits management to ensure that this double counting was avoided, with a clear benefits attribution (more so as rewards were aligned with the achieving of the benefits).

(j) A major challenge for AXN was to manage the utilization of shared resources across the change initiatives. A centralized Resource Management Group (RMG) was formed to track the loading and utilization of these resources. A similar challenge was to identify the major dependencies across change initiatives, especially those cutting across the SBUs. These logical dependencies were a major source of risks.

Whenever there was a mismatch between aggregate demand and supply, the RMG was involved in either developing the skills of the existing resources or using subcontractors from third parties. The RMG reviewed the resource schedule on an ongoing basis to take corrective and predictive actions, which was a major facilitator for effective portfolio implementation.

(k) The Senior Director of Strategy was a key player in ensuring the success of the portfolio by actively participating in the review meetings, communicating the portfolio vision to other Senior Directors, and obtaining the buy-in from other concerned stakeholders. Corporate communication media was also actively involved here, disseminating the progress of change initiatives to concerned stakeholders through diverse methods.

In the chapter on Change Management (Chapter 6), we saw that a Line-Level Manager expects the messages concerning the change to be communicated by the senior-most management. The impact of such messages becomes greater if they also come from their immediate reporting manager. At AXN, these messages were tailored to the audience level for consistency, to ensure their credibility. Multiple media channels such as Corporate brochures, town hall meetings, videos, etc. were used extensively to propagate the current status of the change initiatives, when the benefits were expected to accrue, and what change was needed next to realize these benefits. Such communications raised the overall awareness of the change initiatives across AXN.

(l) As part of the assessment of the portfolio impact, the metrics defined during portfolio definition were monitored rigorously. These metrics included—among others—how much of the money was spent in multiple portfolio segments, the extent of the assessment of slippages in change initiatives, resource utilization rates, the extent of benefits realization, etc. The Senior Director of Strategy also commissioned external reviewers to ensure unbiasedness in checking these values.

9.8 Program Management—Execution

As part of the portfolio, multiple programs were launched in the three SBUs. We provide herein a representative program—how it was initiated, planned, launched, and closed—as an illustration. The program illustrated pertained to the Telecom SBU, whose main client was a major telecom company, which we will call "Airwaves."

Airwaves was expanding into multiple countries and was providing a wallet service for the citizens of these countries, enabling financial transactions (mainly money transfer and receipt) to be processed through mobile services.

Airwaves wanted to have a lower "churn" rate of its customers as a key benefit of its program. Since the Telecom SBU of AXN had been working with Airwaves as a preferred vendor, Airwaves contracted out this program to the AXN Telecom SBU.

Given the benefit statement of Airwaves, the Telecom SBU worked backwards and determined what outcomes needed to be in place to obtain these benefits. The target operating model was derived from the client's perspective. As noted in the chapter on Program Management (Chapter 3), this target operating model had multifarious perspectives. These covered the redesigned aspects outlined below:

(a) *Processes:* These included customer acquisition, registration, procedures for money transfer and receipt as a part of wallet operations, and compliance with local regulations for Airwaves.
(b) *Functional organization structure:* This comprised the recruitment/appointment of area and country managers for Airwaves, their skill sets upgrades, and delineation of organizational reporting structures.
(c) *Technology aspects:* These aspects included redesigning software and middleware applications to ensure the secure processing of client transactions and to secure data storage and maintenance services in "cloud" platforms for Airwaves.
(d) *Information and management dashboards:* These needed to be accessed by Airwaves management, including statistics on user registrations, revenue generated by service lines/geographies, and congestion in network traffic.

Once there was agreement about the target operating model, the Telecom SBU identified projects that needed to be executed to produce the desired outputs. This work mapped to the "component initiation, oversight, and integration" sub-phase as a part of the program execution phase (Chapter 3: Program Management). Once the project outputs were signed off by Airwaves, the Telecom SBU also had the responsibility of transitioning capabilities into Airwaves operational functions—providing training and maintenance support, creating a help desk, and instituting a ticket management system.

A few other programs AXN executed (especially for its shared functions) interfaced with its internal operations. For example, AXN also ran several programs for internal process improvement and HR skill set upgrades to serve its clients better. Such programs were mainly linked to the bottom two perspectives of the balanced scorecard. Whereas the external programs were executed on the basis of contracts with the clients, internal programs were completed based on service-level agreements (SLAs) with concerned functional departments. Internal Change Agents were instrumental in driving change in such cases. Eliminating the redundant processes also resulted in cutting down waste and unessential expenditures.

9.9 Program Execution—Interfacing with Project Management

The concerned Program Managers decided which projects should be taken up to achieve the desired outcomes. For instance, the following are some of the projects implemented by the Electronics Control SBU as part of the program.

(a) Introduction of a new chip design with advanced features at a lower cost
(b) Integration of an existing control system into new car models
(c) Construction of a new facility to manufacture chips referred to in (a) above

The concerned Program Manager was consulted and was actively engaged in selecting the Project Managers for these three projects. During the commencement of the chip design project, the following information was passed on by the Program Manager to its Project Manager (after formal meetings and agreements):

- What are the design features of the new chip? Which are mandatory and non-mandatory features to be addressed as a part of scope prioritization?
- How does this project interface with the other projects in the program? In addition, which are the dependencies with projects in other programs?
- When will this project be launched? What are the schedules and budgets?
- What types of skill sets will be used in the project? How will these be resourced?

- Which are the constraints and risks to be addressed by the project?
- What are the stakeholder concerns to be addressed?
- What standards are to be followed for quality, risk, and configuration management by the project?
- What are the escalation and reporting procedures?

The Program Manager also coordinated the reports from multiple projects and then designed an integrated dashboard report (in collaboration with the SMO) to be sent to the Program Governance Board. The Program Manager was also responsible for dealing with escalations from the projects (mainly risks and issues) and taking appropriate action. The RMG was responsible for allocating resources across multiple projects. There was a tight interface between projects (a) and (c) above, which was handled at the program level.

The Program Governance Board reviewed the overall program progress during phase-gate reviews and authorized the commencement of new projects as per the program management plan.

When each project was closed, the Program Manager took reverse feedback from the concerned Project Manager on lessons learned. This feedback was factored in during further program planning.

During transition management, the Program Manager for the Electronics Control SBU played an active role in facilitating it by overseeing the creation of help desks and user manuals. However, it was the responsibility of the concerned SBU Functional Managers to manage the transition and ensure that the changes were embedded in operations. These Functional Managers were also responsible for the measurement of benefits once the outcomes were stabilized, as per the steps described in the program benefits realization sub-phase.

9.10 Program Closure

When all three projects in the Electronics Control SBU were executed, transitioned, and the outcomes stabilized, the program itself came to a closure. The Program Manager obtained final sign-off for the program from the Functional Manager of the Electronics Control SBU. Again, the lessons learned were discussed retrospectively, and updates were provided to the knowledge database, which was facilitated by the Program

Management Office. All the resources were released, and the RMG updated their availability status. Existing contracts were passed onto the Functional Managers, and the program management team was disbanded. The SMO was responsible for getting the feedback from the Program Manager to the portfolio.

9.11 How Projects in AXN Were Managed—Salient Points

The programs in the portfolio consisted of multiple projects, each following a different lifecycle. Some were engineering-related projects, whereas a few were concerned with process improvement or IT. A Center of Excellence (COE) under the SMO was set up to determine the best practices, which could be applicable to different kinds of projects. Each of the Project Managers consulted this COE database before deciding whether to customize the methodology for their respective projects. However, it was noted that a few projects were started with minimal reference to the common databases of best practices and metrics. In this way, it could be stated that the project management maturity for AXN was in between Levels 2 and 3 (as per the maturity definitions given in Chapter 8).

Each of the projects had a Project Support Office to render administrative support to the respective Project Managers. These support offices took care of master database maintenance, configuration management, and coordination with the Team Managers for team progress tracking.

During the user requirements elicitation, the Functional Manager's representatives were involved with the Team Managers to ensure that "user stories" were properly formulated. The effort estimation for the defined scope was done by consulting with the team. The Project Manager received approval for the project management plan (which included an integrated baseline, with scope/schedule and budget, addressing the "triple triangle") from the concerned Program Manager. Many of the Telecom SBU projects followed the Agile methodology for software development, which was imperative in light of their rapidly changing requirements.

The RMG handled resource allocation for the projects. The Project Manager kept the Program Manager and the Project Review Board informed of the progress and received approval before going ahead to the next stage.

The Project Manager was responsible for ensuring the quality of the deliverables in the project by incorporating the applicable quality standards into the team deliverables. In addition, the Project Manager ensured that the deliverables were acceptable through the performance of the User Acceptance Testing by the concerned Functional Managers, before handing over the outputs to the various functional units under the Program Manager's supervision.

9.12 How AXN Enhanced Project Management Competency

Initially, the PM competency was low in AXN, with different SBUs having local systems and procedures for project management. The SMO then took upon itself the responsibilities of the Portfolio Office, which in turn set up the COE. The COE was vested with the following responsibilities:

- To understand the business change lifecycle of various SBUs and design appropriate project and program management standards, including processes to be followed and governance processes to be used
- To impart trainings in project and program management to selected stakeholders
- To create management dashboards for project and program management progress tracking and collation

The COE was also entrusted with the responsibility of selecting an appropriate project/program management information system (PMIS) tool for AXN. The COE considered the following factors before selecting the appropriate tool:

- The suitability of the PMIS tool for AXN. Most of the tools had sophisticated features that were not relevant for AXN. The COE had to ensure that it only paid for the functionalities that AXN could readily use, while keeping the provision open for upward compatibility.
- The COE had to factor in the maturity of AXN while selecting the PMIS tool. For instance, some of the advanced tools supported

features such as Monte-Carlo simulation analysis, which were not suitable for AXN at this time and were not considered for initial deployment.
- This tool was implemented first in the Financial Services division, as it had more PM maturity. Once it got stabilized, it was rolled out incrementally to other SBUs. In this way, earlier lessons learned were able to be subsequently factored into the implementation.

The COE and the SMO also commissioned an external agency to assess the current PM maturity of AXN. It also corroborated that different SBUs were having varying levels of PM maturity—for instance, the Financial Services SBU had the highest level of PM maturity, followed by the Telecom SBU, and, lastly, by the Electronics Control SBU. This assessment mirrored the extent and intensity of the projects undertaken by various SBUs, and guided where the improvements should next be implemented.

9.13 Portfolio Management Implementation—A Retrospective

The initiatives launched by AXN Corporation were rolled out in phases across a duration of about 14 months. Initially, there was a huge skepticism among some of the Functional Managers on its success, as they had seen many change initiatives being launched with "big fanfare" and then floundering. The sustained management commitment, championed by the CEO, made all the difference in its success this time. By the end of third year of the implementation, revenues went up by 60% from the base level before the launch of the program, and the target of doubling revenues in five years seemed to be on track. Profitability rose to 20% during these three years.

More importantly, improving top-line and bottom-line figures restored confidence in the staff, reducing attrition and increasing staff morale. Aspects that required improvement included the enhancement of PM competency uniformly across all the SBUs and better alignment of rewards and recognition of achievements. These gaps were being worked on by the SMO and the Director of Human Resources, respectively.

Overall, the AXN senior management rated the success of new initiatives in the portfolio as "above average." More importantly, this implementation gave confidence to the SMO for taking more complex change initiatives in alignment with AXN's Corporate strategy.

References

1. Porter, M. E. (1980). *Competitive Strategy: Techniques for Analyzing Industries and Competitors.* New York: Free Press.

Appendix: Structure of Major Portfolio, Program, and Project Artifacts

Note: The following are specimen templates, which are broadly based on global best practices. Practitioners need to customize these templates based on the particular context of project portfolio management at their respective organizations.

1. Portfolio Governance Framework

Purpose/Description	Typical Contents
Purpose: To provide definitive guidance to all stakeholders on the portfolio management practices and procedures adopted by the organization. This guidance is prepared at the organizational level and is owned by the Portfolio Steering Group (PSG).	Description of the organizational vision, mission, and strategic objectives.
	Objectives of portfolio management in the organization.
	Description of how portfolio definition, prioritization, and implementation will occur in the organization. (This can also refer to the business change lifecycle of the organization and the criteria used for portfolio segmentation, prioritization, and balancing.)
	Description of major roles and responsibilities in the portfolio management lifecycle, including guidelines for decision making.
	Description of how the portfolio governance framework will be integrated with the organizational governance.

2. Portfolio Implementation Plan

Purpose/Description	Typical Contents
Purpose: To provide a baseline against which the portfolio progress will be monitored via the portfolio dashboard. This plan also contains reference to various delivery plans, such as the portfolio governance plan, the portfolio risk management plan, etc., to guide the Portfolio Manager during portfolio implementation. This document can be drafted by the Portfolio Manager and other Subject Matter Experts (SMEs), with the assistance of the Portfolio Office. The document needs to be approved by the PSG and is used for portfolio progress tracking by the Portfolio Progress Monitoring Group (PMG).	Description of the portfolio and its composition.
	Description of the benefits to be obtained by implementing the portfolio and how they link up with the organizational strategic objectives.
	Associated scope, schedule, and budgets for the portfolio—delivery plan and associated timelines.
	Portfolio governance plan.
	Portfolio risk management plan.
	Portfolio stakeholder engagement and communications management plans.
	Portfolio progress reporting systems, including the portfolio dashboard format.
	Roles and responsibilities of various stakeholders.

3. Program Mandate

Purpose/Description	**Typical Contents**
Purpose: To provide an external trigger for a program. This document is issued by the Portfolio Manager/Senior Management. The program mandate indicates why the current change initiative could be considered to be run as a program and is a trigger for the program initiation phase. Once the program charter is prepared, it becomes a reference document for the program definition phase.	Strategic objectives of the proposed program initiative. This can include reference to any prior "feasibility study" done.
	Summary of the current "as-is" state.
	Explanation of any external/other forces triggering the program.
	Description of what the program is intended to deliver in terms of new services and operational capabilities/ benefits and outcomes envisaged/ program success criteria.
	Name of proposed Program Sponsor.
	Any current or anticipated initiatives to be included in the program (for emergent programs).
	Expectations of time/cost/ constraints in which the program will operate.
	"High-level" business case for the program.
	Description of initial assurance arrangements.

4. Program Charter

Purpose/Description	Typical Contents
Purpose: To formally authorize the commencement of program definition. The program charter is the first formal document prepared as a part of the program lifecycle during the program initiation phase. This document can be drafted by the prospective Program Manager and needs the approval of the Program Sponsor and the Portfolio Steering Group (PSG).	Name of the program.
	Names and contact information for the Program Manager/Program Sponsor.
	Description of the strategic alignment of the program with the organizational objectives and the portfolio.
	The program outline vision statement.
	The program outline business case.
	Benefits envisaged to be obtained.
	Constraints/assumptions.
	Macro-level scope/estimated costs/efforts/timescales/initial high-level roadmap.
	Major program-level risks and a description of how they are proposed to be addressed.
	Recommended program governance structure.
	Extent of stakeholder support/initial stakeholder considerations.
	Initial list of component projects and their business cases (if applicable).
	Program success criteria.

5. Program Scope Baseline

Purpose/Description	Typical Contents
Purpose: To provide a baseline against which the program's success can be measured. The program scope baseline is finalized during the program definition phase. It includes the scope statement, program Work Breakdown Structure (WBS), and the program success criteria. The Program Manager is the document owner for the baseline document.	Program name. Program vision statement/program benefits expected. Program deliverables/initial descriptions of the program components/program-level WBS. Program boundaries (interfaces with other programs or projects). Assumptions and constraints. Program criteria for success.

6. Program Benefits Realization Plan

Purpose/Description	Typical Contents
Purpose: To provide a baseline schedule outlining when the program benefits are expected to be realized, how they will be measured, and who will be the benefit owners. This document is initially prepared during the program definition phase by the Program Manager, with inputs from the Functional Managers. It is updated during the program lifecycle, and during the program closure phase it is handed over to the Functional Managers.	Program name.
	List of benefits the program is expected to achieve and the measurement criteria to be used.
	Timeline showing when the benefits/ outcomes are likely to be achieved.
	Milestones for program benefits review.
	Dependencies across benefits.
	Benefits map showing the link to strategic objectives and the components needed to realize the benefits.
	Benefit owners and description of how they are expected to measure and sustain the benefits.
	Disbenefits that are expected.
	Transition plan.
	Details on how this plan will be maintained post-program closure (benefit sustainment plan).

7. Program Benefit Card

Purpose/Description	Typical Contents
Purpose: To summarize the basic information concerning a benefit as a single reference document. A benefit card captures all the relevant features of a benefit. Typically, the benefit card is owned by the concerned Functional Manager, in whose functional area the benefits will be realized. The benefit card also outlives the program lifecycle and is handed over to the benefit owner during the program closure phase.	Benefit identifier number.
	Description of the benefit.
	Current baseline measurements of the benefit
	Expected improvements—with timescales for the benefit.
	Details of the facilitating program, projects, operations, and outcomes required to attain the benefit.
	Cost of obtaining this benefit—derived from corresponding projects and operations costs.
	How, by whom, and when the benefit will be measured.
	Benefit owner/recipient/organizational structure needed to sustain the benefit.
	Major risks concerning the achievement of the benefit and suggested risk responses.

8. Program Benefits Management Strategy

Purpose/Description	Typical Contents
Purpose: To give guidance on the benefits management lifecycle during the program. The benefits management strategy is prepared by the Program Manager, with inputs from the Portfolio Manager and the Functional Managers during the program definition phase. This document gets updated during the program lifecycle. The benefits management strategy also outlives the program lifecycle and is handed over to the portfolio management during the program closure phase.	Program name.
	Guidance on how benefits are identified, classified, prioritized, and measured during the program lifecycle and after its closure.
	Metrics used for benefits measurement.
	Description of the functional areas impacted by the benefits from the program.
	Guidelines on the events at which the benefits will be measured.
	Description of any tools and techniques that could be used to measure the benefits.
	Guidance on roles and responsibilities for the benefits management lifecycle.

9. Program Communications Management Plan

Purpose/Description	Typical Contents
Purpose: To define a schedule of communication-related events in a program environment. The program communications management plan is prepared by the Program Manager during the program definition phase and updated during the program lifecycle. It is a component of the overall program management plan.	Program name.
	List of stakeholders and their communication requirements.
	Information to be communicated to stakeholders—including their content, structure, periodicity, etc.
	People responsible for communicating and receiving the information; feedback mechanisms.
	Technologies to be used for communication and information storage systems.
	Escalation processes.
	Methods to update and refine the communications management plan.
	Metrics to assess the success of the program-level communications.

Note: This plan can be integrated with the program stakeholder engagement plan, which shows the classification of various stakeholders, how they need to be engaged, and how to assess the involvement of the stakeholders during the program lifecycle.

10. Program Risk Management Plan

Purpose/Description	Typical Contents
Purpose: The program risk management plan provides guidance on how to manage risks during the program lifecycle. It is created during the program definition phase and is maintained throughout the program lifecycle by the Program Manager. The program risk management plan is to be compliant/aligned with the Corporate risk management strategy.	Program name.
	Details of the Corporate or other industry standards the program risk management plan is required to be compliant with.
	Description of the program risk management process, including explanation of risk identification/risk prioritization /risk response planning/monitoring and controlling procedures.
	Roles and responsibilities for risk management in the organization and in the program.
	Cost and expenditure profiles to manage the risk management process in the program.
	Schedule of risk management actions, including the communication of risk status.
	Templates for the program risk register to follow for the program.

11. Program Risk Register

(This register comprises information that applies to all the risks in the program.)

Purpose/Description	Typical Contents
Purpose: The program risk register is a central repository for all the risks relevant for the program. It is created during the program definition phase and updated throughout the program lifecycle. Although the Program Manager is the document owner for this register, the Program Manager needs to take the inputs from the risk owners, etc., while updating the risk register. The risk register gets closed during the program closure phase and pending risks get transferred to concerned risk owners.	Program name.
	Risk identifier/description.
	Root causes of the risk.
	Likely probability, impact, and proximity of the risk /expected monetary value of the risk.
	Risk author/owner and the responsibilities assigned to them.
	Early warning identifiers of the risk.
	Agreed-upon risk responses.
	Budget and scheduled activities to implement the chosen option for the risk.
	Residual status of the risk after planned responses have been taken.
	Any secondary risks.
	Current status of the risk (monitored at different dates).

12. Program Financial Management Plan

<table>
<tr><th>Purpose/Description</th><th>Typical Contents</th></tr>
<tr><td rowspan="8">Purpose: The program financial management plan provides guidance on how to manage the program finances.

It is created during the program definition phase and updated throughout the program lifecycle.

Although the Program Manager is the document owner, the Program Manager needs to take the inputs from diverse stakeholders, especially the finance department.</td><td>Program name.</td></tr>
<tr><td>Program financial framework.</td></tr>
<tr><td>Program funding schedules/ milestones.</td></tr>
<tr><td>Baseline budget.</td></tr>
<tr><td>Component payment schedules/ contractor payment schedules/ funding milestone information.</td></tr>
<tr><td>Financial reporting processes/ accounting practices to be used.</td></tr>
<tr><td>Compliance and regulatory-related issues/financial metrics to be used.</td></tr>
<tr><td>Financial governance-related processes.</td></tr>
</table>

13. Program Quality Management Plan

Purpose/Description	Typical Contents
Purpose: The program quality management plan sets guidelines for implementing the quality management lifecycle in the program. It is created during the program definition phase and updated throughout the program lifecycle. Although the Program Manager is the document owner, the Program Manager needs to take the inputs from diverse stakeholders, especially the Corporate quality department.	Program name.
	Corporate or other required industry standards to be adopted for the program quality systems.
	Approach to be used for quality planning, quality assurance, and quality control in the program.
	Minimum set of quality standards that will be applied to component deliverables.
	Schedule of quality management actions, including assurance reviews and health check assessments.
	Description of roles and responsibilities for program quality functions.
	Information and resource requirements to support quality-related actions.
	Procedures and tools to be used for quality-related actions.
	Cost and expenditure profiles to manage quality actions.

14. Program Resource Management Plan

Purpose/Description	Typical Contents
Purpose: The program resource management plan provides guidelines on how to manage resources (including human and non-human resources) in the program lifecycle. It is created during the program definition phase and updated throughout the program lifecycle. Although the Program Manager is the document owner, the Program Manager needs to take into consideration the input of diverse stakeholders, especially that of the human resources department.	Program name.
	List of resources (including people/facilities/office space, etc.) needed in the program and when they will be required.
	Details of how the resource blending between internal and external resources will be done.
	Likely costs of acquiring these resources.
	Explanation of how these resources will be managed, including treatment of shared resources, disposition, and reintegration into the organization.
	Roles and responsibilities for resource management.

15. Program Component List

Purpose/Description	Typical Contents
Purpose: This list is a register of all the components in the program and is used to give a summary snapshot of various components. It is prepared by the Program Manager during the program definition phase and updated during the execution phase in the program lifecycle.	List of the components in the program.
	Outline information on the scope, time, and cost of each of the components and their resource/quality requirements.
	Description of the component interdependencies.
	Links showing the contribution of each component to the program's benefits.
	Any other information relating to the component, such as the details of any specific standards to be used for its execution.

Note: The program roadmap will also contain a summary of references to various components in a timeline.

16. Program Transition Plan

Purpose/Description	Typical Contents
Purpose: The program transition plan provides guidelines on how to manage the transition successfully. It is prepared by the Program Manager during the program definition phase, with inputs from the concerned Functional Managers. This document is updated during the program execution phase, especially as components go live for transition.	Program name.
	Pre-transition step: Schedule and details of activities to be done during this step, including benefits baseline values establishment, preparing the impacted areas for change, designing the training and handover plan, developing backup and rollback plans, dates during which user acceptance /integration testing, etc., will be done for the component deliverables.
	Transition step: Schedule and details of activities for migration/cutover to operations, trainings, parallel runs, and handover.
	Post-transition step: Schedule and details of activities for outcome stabilization and benefits measurement.

Note: The above information will be noted for each of the transitions in the program.

17. Program Governance Plan

Purpose/Description	Typical Contents
Purpose: The program governance plan is a high-level guidance document informing how the program will be monitored and controlled. Though this plan is prepared by the Program Manager (during the program definition phase), it needs to be vetted specifically by the Program Governance Board before its approval. The governance plan will be updated during the program lifecycle. This plan is used both by the Program Manager and the Program Governance Board for their work.	Program name.
	Program goals summary.
	Structure and composition of the Program Governance Board.
	Individual roles and responsibilities of the Program Governance Board members.
	Frequency and purpose of governance board meetings, including communication and information requirements.
	Planned phase gate/assurance/ milestone reviews, including their expected schedule and the goals of these meetings.
	Component initiation/transition/ closure criteria.
	Issue escalation processes from the Program Manager to the Program Governance Board.
	Issue escalation processes from the Component Manager to the Program Manager.
	Criteria by which the portfolio will oversee the program progress.
	Program success and closure criteria.
	Details of how the program will govern its components.

18. Program Management Plan

Purpose/Description	Typical Contents
Purpose: The program management plan serves as the master document for the Program Manager to use as a reference to plan and control the program lifecycle. This plan is prepared by the Program Manager during the program definition phase and updated during the program execution phase. It is mainly consulted during the program closure phase.	Program name.
	Alignment of the program with the strategic objectives.
	All program subsidiary plans, including: • Scope/schedule/cost management plans • Resource/procurement management plans • Program governance plan • Program stakeholder engagement plan • Program financial management plan • Program risk and issue management plans • Stakeholder engagement and Communications management plans • Benefits realization plan • Program quality management plan
	Program baselines—covering scope/schedule/budgets.
	Program roadmap/list of benefits the program is expected to obtain.
	Transition plan.
	Description of how program–component interfaces will be maintained.
	List of components/projects.

Note: The above plan is updated during the program lifecycle, along with the program business case and target operating model (as required).

19. Program Target Operating Model

<table>
<tr><th>Purpose/Description</th><th>Typical Contents</th></tr>
<tr><td rowspan="3">Purpose: The target operating model informs the organizational structure, processes, and business functions of impacted functional departments.

The gap between the "as-is" and "to-be" target operating model indicates what needs to be addressed by the projects, transitions, and outcome management to reach the "to-be" state.

This document is prepared by the Program Manager during the program definition phase, and is used during the program execution phase.

For multiple iterations of the program execution phase, the target operating model could be incrementally defined.

Once the final target is reached, the program can close.</td><td>Name of the program.</td></tr>
<tr><td>Summary of "as-is" state.</td></tr>
<tr><td>Details of the "to-be" state of the organization. This can include following information for the impacted functional departments:

• Roles and responsibilities, skill sets, staffing levels, reporting structure, style of functioning
• Processes, work flow changes, performance levels required
• Technology changes (including redesigned IT support), tools, infrastructure support, etc.
• Redesigned management dashboards, information flows, and statistics required to support the future business operations and decision making</td></tr>
</table>

20. End Program Report

Purpose/Description	Typical Contents
Purpose: The end program report is the last formal document produced during the program lifecycle, leading to the winding down of the program. The end program report is produced by the Program Manager during the program closure phase. This report is approved by the Program Governance Board (and Portfolio Steering Group, if needed). Thereafter, the program can formally close. The report is also produced for abnormal closure of the program. In this case, the contents (noted in the right-hand column) are modified appropriately, depending on when the program is prematurely closed.	Program name.
	Program Manager's assessment on how the program went.
	Confirmation that all the components have been transitioned and outcomes realized.
	Linkage with benefits realization plan—stating the extent of benefits realized and when the benefits are likely to accrue.
	Details of the arrangements for transfer of ongoing contracts to the concerned Functional Managers.
	Arrangements for transferring ownership of pending risks and issues.
	Program lessons learned report and feedback to Corporate strategy.
	Program Manager's assessment of the team performance and recommendations for skill improvements.
	Program wind-down plan, including resource disposition and program assets archival plan.

21. Project Charter

Purpose/Description	Typical Contents
Purpose: The project charter is the "kick-off" document in the project lifecycle. It is drafted by the Project Manager during the project initiating process and gets approved by the Project Sponsor. The charter becomes a key document for finalising the project management plan during the project setup process.	Project name.
	Names and contact information of the Project Manager and the Project Sponsor.
	Project objectives (scope, schedule, budget, milestones, project-level tolerance information).
	List of major project stakeholders.
	Constraints/assumptions/ dependencies with other change initiatives in the program/ organization.
	Project outline business case.
	Expected major risks and issues.
	References to governance arrangements (e.g., quality, risk, and communications), including alignment with program governance standards.
	Project progress reporting and escalation guidelines.
	References to contracts (in case of external projects) and feasibility studies (for internal projects).

22. Project Business Case

Purpose/Description	Typical Contents
Purpose: The project business case is used to assess the initial and ongoing viability of the project. The outline business case is prepared during the project initiating process and gets detailed during the setup process. Whenever issues and risks arise, the impact analysis will focus on project viability. The Project Sponsor becomes the owner of the project business case. The detailed business case gets updated during various stages of the project.	Project name.
	Rationale for the project.
	Project alignment with the program/organizational objectives.
	Project scope/schedule overview.
	Description of project delivery options considered and, for each option, the cost–benefit analysis.
	Details of the selected project delivery option.
	Expected benefits.
	Major expected risks and the risk responses.

23. Project Scope Management Plan

Purpose/Description	Typical Contents
Purpose: The project scope management plan is a guidance document on how to manage the scope in a project. It can include the requirements management plan. This document is prepared by the Project Manager during the project setup process and becomes integrated into the project management plan.	Project name.
	Details of the scope management process.
	Details of the requirements management process in the project.
	Description of how the project WBS will be created and maintained.
	Details of how the scope baseline will be created and updated.
	Details of how scope control will be managed in the project.

24. Project Cost Management Plan

Purpose/Description	Typical Contents
Purpose: The project cost management plan is a guidance document on how to manage the costs/expenditures in a project. This document is prepared by the Project Manager during the project setup process and becomes integrated into the project management plan.	Project name.
	Details of the cost and expenditure management process in the project.
	Details on how to estimate costs for the activities and roll up to the project level.
	Details of project cost accounting/ updating and reporting procedures.
	Details of how the cost baseline will be created and updated.
	Details of cost-control processes.

25. Project Quality Management Plan

Purpose/Description	Typical Contents
Purpose: The project quality management plan provides guidance on how to manage quality in the project environment. As in other management plans, this document is prepared by the Project Manager during the project setup process and becomes integrated into the project management plan.	Project name.
	Project scope description.
	Description of the organizational quality policy and how it needs to be tailored for the project.
	Quality standards/test criteria to be adopted for the project deliverables.
	Metrics and checklists to be used to enforce quality.
	Tools and techniques to be used to ensure quality planning, assurance, and control.
	Schedule of project quality responsibilities.
	Details of project quality records to be maintained.
	Project quality team responsibilities.

26. Project Resource Management Plan

Purpose/Description	Typical Contents
Purpose: The project resource management plan provides guidance on how to manage resources in the project environment. As in other management plans, this document is prepared by the Project Manager during the project setup process and becomes integrated into the project management plan.	Project name.
	Description of the resource pool from which the resources would be acquired.
	Skill sets required for the project (stage-wise description).
	Staffing management plan, including details about when the resources will be needed for the project and when they will be released.

27. Configuration Item Record

Purpose/Description	Typical Contents
Purpose: The configuration item record is produced for each product in the project and provides a complete snapshot of the current status of the product. The Project Management Office can support the Project Manager in the creation and updating of the configuration item record.	Project name.
	Product (Item) code/name as assigned by the configuration management system.
	Description of the product.
	Relationship with other products.
	Last date of updating.
	Current status of the product (e.g., under development/tested/accepted, etc.) as of the last update.
	Product owner.

28. Project Management Plan

Purpose/Description	Typical Contents
Purpose: The project management plan is a master reference document for the project. This document is prepared by the Project Manager during the project setup process and gets continuously updated during the project lifecycle.	Project name.
	Name and contact information for the Project Manager/Sponsor.
	Project WBS/WBS dictionary/scope statement.
	Project schedule baseline.
	Cost baseline.
	Subsidiary plans including the following plans : • Scope management plan • Schedule management plan • Cost management plan • Quality management plan • Resource management/staffing management plan • Stakeholder engagement plan/ communications management plan • Risk management plan • Procurement management plan • Project governance plan

29. Team Progress Report

Purpose/Description	Typical Contents
Purpose: The team progress report updates the Project Manager on the status of the work allocated to the team. It is prepared by the Team Manager and is sent to the Project Manager as per defined periodicities.	Team Manager name.
	Report submission date.
	Deliverable identification/name (for various deliverables in the report).
	Planned and actual start dates of the deliverable.
	Planned and expected end dates of the deliverable.
	Details of work completed during the reporting period and particulars regarding ongoing work .
	Resource utilization trends/effort spent trends.

30. Project Progress Report

Purpose/Description	Typical Contents
Purpose: The project progress report updates the Project Review Board on the status of the project. It is prepared by the Project Manager during the project delivery process and is sent to the Project Review Board, as per defined periodicities.	Report date.
	Status of pending actions covered from the last report.
	Current reporting period summary status (including deliverables completed, schedule status, budget and resource usage, etc.).
	Status of issues and risks.
	Major stakeholders—pending issues.

31. End Project Report

Purpose/Description	Typical Contents
Purpose: The end project report is the last formal report produced during the project lifecycle and signifies that the project is about to close. This report is produced by the Project Manager during the project closing process of the last stage of the project. At the end of each stage, an "end stage report" is produced by the Project Manager with contents quite similar to the "end project report," covering the details for this stage. The end project report can also be produced for abnormally terminating projects. In this context, the contents in the right-hand column are modified appropriately, depending on when the project was abnormally closed.	Description of the viability of the project business case during the project closure.
	Original and updated records of project management baselines.
	Review of the project objectives and whether they were met successfully.
	Acceptance status of various deliverables, including any deviations.
	Sign-off of records of final deliverables.
	Handover of records to functional departments, including actions proposed for pending issues and risks.
	Review of the team performance.
	Project Manager's assessment of what went right and what could be improved.
	Lessons learned and feedback to program management/portfolio management.
	Project wind-down arrangements (including disposition of resources).
	Project deliverable archival arrangements.

Glossary

Chapter 1

CEO	Chief Executive Officer
IT	Information Technology
P3M	Portfolio, Programme and Project Management
PESTLE	Political, Economic, Social, Technological, Legal, Environmental
PMI	Project Management Institute

Chapter 2

AHP	Analytic Hierarchy Process
ARR	Accounting Rate of Return
BAU	Business As Usual
BCG	Boston Consulting Group
BSC	Balanced Scorecard
CEO	Chief Executive Officer
COE	Center of Excellence
CSAT	Customer Satisfaction
CSR	Corporate Social Responsibilities
ERP	Enterprise Resource Planning
HR	Human Resources
IRR	Internal Rate of Return
ISO	International Standards Organization

IT	Information Technology
KPI	Key Performance Indicator
NPV	Net Present Value
PfO	Portfolio Office
PMG	Portfolio Progress Monitoring Group
PSG	Portfolio Steering Group
SBU	Strategic Business Unit
SWOT	Strengths, Weaknesses, Opportunities, and Threats

Chapter 3

BAU	Business As Usual
COE	Center of Excellence
EPM	Enterprise Program/Project Management
ERP	Enterprise Resource Planning
HR	Human Resources
IT	Information Technology
PBS	Product Breakdown Structure
PfO	Portfolio Office
PgMO	Program Management Office
PgWBS	Program Work Breakdown Structure
PI matrix	Probability Impact matrix
PjWBS	Project Work Breakdown Structure
PMG	Portfolio Progress Monitoring Group
PSG	Portfolio Steering Group
RMG	Resource Management Group
SKU	Stock Keeping Unit
SME	Subject Matter Expert
UAT	User Acceptance testing
WBS	Work Breakdown Structure

Chapter 4

CIR	Configuration Item Record
COE	Center of Excellence
EF	Early Finish

ES	Early Start
HR	Human Resources
IRR	Internal Rate of Return
IT	Information Technology
KMO	Knowledge Management Office
LF	Late Finish
LS	Late Start
MoSCow	"Must be done, Should be done, Could be done, Will not be done"
NPV	Net Present Value
OBC	Outline Business Case
PBS	Product Breakdown Structure
PgWBS	Program Work Breakdown Structure
PI	Probability Impact
PjWBS	Project Work Breakdown Structure
PMBOK	Project Management Body of Knowledge
PMI	Project Management Institute
PRB	Project Review Board
R&D	Research and Development
RBS	Risk Breakdown Structure
RCA	Root Cause Analysis
RMG	Resource Management Group
SLA	Service-Level Agreement
SME	Subject Matter Expert

Chapter 5

BAU	Business As Usual
IT	Information Technology
PMG	Portfolio Progress Monitoring Group
SLA	Service-Level Agreement

Chapter 6

CPIG	Customers, Providers, Influencers, Governance
PjMO	Project Management Office

PMIS	Project Management Information System(s)
PMO	Project Management Office
ROI	Return on Investment

Chapter 7

PfO	Portfolio Office
PSG	Portfolio Steering Group

Chapter 8

BAU	Business As Usual
CEO	Chief Executive Officer
CMM	Capability Maturity Model
COE	Center of Excellence
CPI	Continuous Process Improvement
EPMO	Enterprise Project Management Office
IT	Information Technology
OPM3	Organizational Project Management Maturity Model
P3M	Portfolio, Program and Project Management
P3M3	Portfolio, Programme and Project Management Maturity Model
PDCA	Plan–Do–Check–Act
PfO	Portfolio Office
PMIS	Project/ Program Management Information System
SBU	Strategic Business Unit
SEI	Software Engineering Institute
SME	Subject Matter Expert

Chapter 9

BAU	Business As Usual
BCG	Boston Consulting Group
BOD	Board of Directors
CEO	Chief Executive Officer
CFO	Chief Financial Officer

COE	Center of Excellence
HR	Human Resources
IRR	Internal Rate of Return
IT	Information Technology
KRA	Key Result Area
NPV	Net Present Value
OEM	Original Equipment Manufacturer
PESTLE	Political, Economic, Social, Technological, Legal, Environmental
PfO	Portfolio Office
PMIS	Project Management Information System
PPM	Project and Program Managers
R&D	Research and Development
RMG	Resource Management Group
SBU	Strategic Business Unit
SLA	Service-Level Agreement
SMO	Strategy Management Office
SWOT	Strengths, Weaknesses, Opportunities, and Threats

Suggested Reading

The following Suggested Reading list contains books and articles of interest to readers in project, program, and portfolio management and related fields.

Agile Manifesto Team. (2001). "Manifesto for Agile Software Development." Open source. Available from http://www.agilemanifesto.org/.

AXELOS. (2013). *Portfolio, Programme and Project Offices (P3O®).* The Stationery Office (TSO).

AXELOS. (2010). *Management of Risk: Guidance for Practitioners.* The Stationery Office (TSO).

AXELOS. (2011). *Management of Portfolios (MoP®).* The Stationery Office (TSO).

AXELOS. (2011). *Managing Successful Programmes (MSP®).* The Stationery Office (TSO).

Bridges, W. (2009). *Managing the Transitions: Making the Most of Change,* 3rd ed. Da Capo Press.

Brown, J. T. (2014). *The Handbook of Program Management: How to Facilitate Project Success with Optimal Program Management,* 2nd ed. McGraw Hill.

Cameron, E. and Green, M. (2012). *Making Sense of Change Management—A Complete Guide to the Models, Tools and Techniques of Organizational Change,* 3rd ed. Kogan Page.

Fabozzi, F. and Markowitz, H. M., eds. (2011). *The Theory and Practice of Investment Management: Asset Allocation, Valuation, Portfolio Construction, and Strategies,* 2nd ed. Wiley.

Kotter, J. P., Kim, W. C., and Mauborgne, R. (2011). *HBR's 10 Must Reads on Change Management.* Harvard Business Review Press.

Hiatt, J. M. (2006). *ADKAR: A Model for Change in Business, Government and Our Community.* Prosci Learning Center Publications.

Hiatt, J. M. and Creasey, T. (2012). *Change Management: The People Side of Change,* 2nd ed. Prosci Learning Center Publications.

Jenner, S. (2014). *Managing Benefits: Optimizing the Return from Investments,* 2nd ed. The Stationery Office (TSO).
Kaplan, R. S. and Norton, D. P. (1996). *The Balanced Scorecard: Translating Strategy into Action.* Harvard Business Review Press.
Kaplan, R. S. and Norton, D. P. (2000). *The Strategy-Focused Organization: How Balanced Scorecard Companies Thrive in the New Business Environment.* Harvard Business Review Press.
Kaplan, R. S. and Norton, D. P. (2008). *The Execution Premium: Linking Strategy to Operations for Competitive Advantage.* Harvard Business Review Press.
Kotter, J. P. (2012). *Leading Change.* Harvard Business Review Press.
Levin, G. (2010). *Interpersonal Skills for Portfolio, Program and Project Managers.* Management Concepts.
Levin, G. and Wyzalek, J., eds. (2014). *Portfolio Management: A Strategic Approach (Best Practices and Advances in Program Management Series).* CRC Press.
Levin, G., ed. (2012). *Program Management: A Life Cycle Approach (Best Practices and Advances in Program Management Series).* CRC Press.
Levine, H. A. (2005). *Project Portfolio Management: A Practical Guide to Selecting Projects, Managing Portfolios, and Maximizing Benefits.* Jossey-Bass.
Morgan, G. (2006). *Images of Organization,* 2nd ed. Sage Publications, Inc.
Nir, M. (2013). *Agile Project Management: The Agile PMO,* Vol. 1. CreateSpace.
Porter, M. E. (1998). *Competitive Strategy: Techniques for Analyzing Industries and Competitors.* New York: Free Press.
Porter, M. E. (1998). *The Competitive Advantage: Creating and Sustaining Superior Performance.* Free Press.
Project Management Institute (PMI). (2013). *Organizational Project Management Maturity Model:(OPM3®),* 3rd ed. PMI.
Project Management Institute. (2013). *A Guide to the Project Management Body of Knowledge: PMBOK® Guide,* 5th ed. PMI.
Project Management Institute. (2013). *The Standard for Program Management,* 3rd ed. PMI.
Ross, J. W., Weill, P., and Robertson, D. C. (2006). *Enterprise Architecture As Strategy: Creating a Foundation for Business Execution.* Harvard Business Review Press.
Schein, E. H. (2010). *Organizational Culture and Leadership,* 4th ed. Jossey-Bass.
Smith, R., King, D., Sidhu, R., and Skelsey, D. (2014). *The Effective Change Manager's Handbook: Essential Guidance to the Change Management Body of Knowledge.* Kogan Page.
Stern, C. W. and Deimler, M. S., eds. (2006). *The Boston Consulting Group on Strategy: Classic Concepts and New Perspectives,* 2nd ed. Wiley.
The Change Management Institute. (2014). *The Effective Change Manager: The Change Management Body of Knowledge.* Vivid Publishing.
The Stationery Office. (2009). *Managing Successful Projects with PRINCE2®.* TSO.
Wren, D. A. and Bedeian, A. G. (2008). *The Evolution of Management Thought,* 6th ed. Wiley.

Index

C